Everyday Mathematics and Standards for Mathematical Practice

Introduction	**page 2**
***Everyday Mathematics* Goals for Mathematical Practice**	**page 3**
Integrating the Standards for Mathematical Practice into Classroom Instruction	**page 6**
Opportunities in *Everyday Mathematics* for Addressing the Goals for Mathematical Practice	**page 8**
***Everyday Mathematics* Goals for Mathematical Practice (List)**	**page 10**
Grade K Section 1	**page 11**
Grade K Section 2	**page 16**
Grade 1 Unit 1	**page 21**
Grade 1 Unit 2	**page 25**
Grade 2 Unit 1	**page 29**
Grade 2 Unit 2	**page 32**
Grade 3 Unit 1	**page 35**
Grade 3 Unit 2	**page 38**
Grade 4 Unit 1	**page 41**
Grade 4 Unit 2	**page 44**
Grade 5 Unit 1	**page 47**
Grade 5 Unit 2	**page 50**
Grade 6 Unit 1	**page 53**
Grade 6 Unit 2	**page 58**

Introduction

The *Common Core State Standards for Mathematics* (CCSS-M) includes both Content Standards and Standards for Mathematical Practice. The Content Standards specify skills and understandings that students should develop at each grade, whereas the Standards for Mathematical Practice identify "processes and proficiencies" that students should develop throughout their mathematical education at all levels, K–12 (CCSS-M, page 6).

The processes and proficiencies in the CCSS-M Standards for Mathematical Practice are at the core of *Everyday Mathematics*. From the beginning, *Everyday Mathematics* has approached mathematics through applications, modeling, and problem solving. The program has always stressed multiple representations, communication, tools, mathematical reasoning, and making sense of concepts and procedures. Therefore, the eight CCSS-M Standards for Mathematical Practice are inherent in *Everyday Mathematics*.

The CCSS edition of *Everyday Mathematics* has thus given us a welcome opportunity to highlight processes that have always been central to *Everyday Mathematics*. For this edition, we have taken the CCSS-M Standards for Mathematical Practice and operationalized them as explicit goals that fit into our overall goal framework. These *Everyday Mathematics* Goals for Mathematical Practice succinctly convey the key ideas in the CCSS-M Standards for Mathematical Practice. They are designed to help teachers make sense of the CCSS-M practice standards and to better focus their instruction on these important standards.

Everyday Mathematics® Goals for Mathematical Practice

The *Everyday Mathematics* authors have distilled the CCSS-M Standards for Mathematical Practice into a set of 23 "Goals for Mathematical Practice" that are intended to be more usable for elementary school teachers and students. These *Everyday Mathematics* Goals for Mathematical Practice (GMPs) provide a framework for instruction in mathematical practices similar to the framework for mathematical skills and understandings provided by the *Everyday Mathematics* Program Goals and Grade-Level Goals. The chart below provides the full text of each CCSS-M Standard for Mathematical Practice in the left-hand column along with the corresponding GMPs in the right-hand column.

Common Core State Standards for Mathematical Practice	*Everyday Mathematics* Goals for Mathematical Practice
Standard for Mathematical Practice 1: Make sense of problems and persevere in solving them.	
Mathematically proficient students start by explaining to themselves the meaning of a problem and looking for entry points to its solution. They analyze givens, constraints, relationships, and goals. They make conjectures about the form and meaning of the solution and plan a solution pathway rather than simply jumping into a solution attempt. They consider analogous problems, and try special cases and simpler forms of the original problem in order to gain insight into its solution. They monitor and evaluate their progress and change course if necessary. Older students might, depending on the context of the problem, transform algebraic expressions or change the viewing window on their graphing calculator to get the information they need. Mathematically proficient students can explain correspondences between equations, verbal descriptions, tables, and graphs or draw diagrams of important features and relationships, graph data, and search for regularity or trends. Younger students might rely on using concrete objects or pictures to help conceptualize and solve a problem. Mathematically proficient students check their answers to problems using a different method, and they continually ask themselves, "Does this make sense?" They can understand the approaches of others to solving complex problems and identify correspondences between different approaches.	**GMP 1.1** Work to make sense of your problem. **GMP 1.2** Make a plan for solving your problem. **GMP 1.3** Try different approaches when your problem is hard. **GMP 1.4** Solve your problem in more than one way. **GMP 1.5** Check whether your solution makes sense. **GMP 1.6** Connect mathematical ideas and representations to one another.
Standard for Mathematical Practice 2: Reason abstractly and quantitatively.	
Mathematically proficient students make sense of quantities and their relationships in problem situations. They bring two complementary abilities to bear on problems involving quantitative relationships: the ability to *decontextualize*—to abstract a given situation and represent it symbolically and manipulate the representing symbols as if they have a life of their own, without necessarily attending to their referents—and the ability to *contextualize,* to pause as needed during the manipulation process in order to probe into the referents for the symbols involved. Quantitative reasoning entails habits of creating a coherent representation of the problem at hand; considering the units involved; attending to the meaning of quantities, not just how to compute them; and knowing and flexibly using different properties of operations and objects.	**GMP 2.1** Represent problems and situations mathematically with numbers, words, pictures, symbols, gestures, tables, graphs, and concrete objects. **GMP 2.2** Explain the meanings of the numbers, words, pictures, symbols, gestures, tables, graphs, and concrete objects you and others use.

© Copyright 2010. National Governors Association Center for Best Practices and Council of Chief State School Officers. All rights reserved.

Common Core State Standards for Mathematical Practice	*Everyday Mathematics* Goals for Mathematical Practice
Standard for Mathematical Practice 3: Construct viable arguments and critique the reasoning of others.	
Mathematically proficient students understand and use stated assumptions, definitions, and previously established results in constructing arguments. They make conjectures and build a logical progression of statements to explore the truth of their conjectures. They are able to analyze situations by breaking them into cases, and can recognize and use counterexamples. They justify their conclusions, communicate them to others, and respond to the arguments of others. They reason inductively about data, making plausible arguments that take into account the context from which the data arose. Mathematically proficient students are also able to compare the effectiveness of two plausible arguments, distinguish correct logic or reasoning from that which is flawed, and—if there is a flaw in an argument—explain what it is. Elementary students can construct arguments using concrete referents such as objects, drawings, diagrams, and actions. Such arguments can make sense and be correct, even though they are not generalized or made formal until later grades. Later, students learn to determine domains to which an argument applies. Students at all grades can listen or read the arguments of others, decide whether they make sense, and ask useful questions to clarify or improve the arguments.	**GMP 3.1** Explain both what to do and why it works. **GMP 3.2** Work to make sense of others' mathematical thinking.
Standard for Mathematical Practice 4: Model with mathematics.	
Mathematically proficient students can apply the mathematics they know to solve problems arising in everyday life, society, and the workplace. In early grades, this might be as simple as writing an addition equation to describe a situation. In middle grades, a student might apply proportional reasoning to plan a school event or analyze a problem in the community. By high school, a student might use geometry to solve a design problem or use a function to describe how one quantity of interest depends on another. Mathematically proficient students who can apply what they know are comfortable making assumptions and approximations to simplify a complicated situation, realizing that these may need revision later. They are able to identify important quantities in a practical situation and map their relationships using such tools as diagrams, two-way tables, graphs, flowcharts and formulas. They can analyze those relationships mathematically to draw conclusions. They routinely interpret their mathematical results in the context of the situation and reflect on whether the results make sense, possibly improving the model if it has not served its purpose.	**GMP 4.1** Apply mathematical ideas to real-world situations. **GMP 4.2** Use mathematical models such as graphs, drawings, tables, symbols, numbers, and diagrams to solve problems.
Standard for Mathematical Practice 5: Use appropriate tools strategically.	
Mathematically proficient students consider the available tools when solving a mathematical problem. These tools might include pencil and paper, concrete models, a ruler, a protractor, a calculator, a spreadsheet, a computer algebra system, a statistical package, or dynamic geometry software. Proficient students are sufficiently familiar with tools appropriate for their grade or course to make sound decisions about when each of these tools might be helpful, recognizing both the insight to be gained and their limitations. For example, mathematically proficient high school students analyze graphs of functions and solutions generated using a graphing calculator. They detect possible errors by strategically using estimation and other mathematical knowledge. When making mathematical models, they know that technology can enable them to visualize the results of varying assumptions, explore consequences, and compare predictions with data. Mathematically proficient students at various grade levels are able to identify relevant external mathematical resources, such as digital content located on a website, and use them to pose or solve problems. They are able to use technological tools to explore and deepen their understanding of concepts.	**GMP 5.1** Choose appropriate tools for your problem. **GMP 5.2** Use mathematical tools correctly and efficiently. **GMP 5.3** Estimate and use what you know to check the answers you find using tools.

Common Core State Standards for Mathematical Practice	Everyday Mathematics Goals for Mathematical Practice
Standard for Mathematical Practice 6: Attend to precision.	
Mathematically proficient students try to communicate precisely to others. They try to use clear definitions in discussion with others and in their own reasoning. They state the meaning of the symbols they choose, including using the equal sign consistently and appropriately. They are careful about specifying units of measure, and labeling axes to clarify the correspondence with quantities in a problem. They calculate accurately and efficiently, express numerical answers with a degree of precision appropriate for the problem context. In the elementary grades, students give carefully formulated explanations to each other. By the time they reach high school they have learned to examine claims and make explicit use of definitions.	**GMP 6.1** Communicate your mathematical thinking clearly and precisely. **GMP 6.2** Use the level of precision you need for your problem. **GMP 6.3** Be accurate when you count, measure, and calculate.
Standard for Mathematical Practice 7: Look for and make use of structure.	
Mathematically proficient students look closely to discern a pattern or structure. Young students, for example, might notice that three and seven more is the same amount as seven and three more, or they may sort a collection of shapes according to how many sides the shapes have. Later, students will see 7×8 equals the well remembered $7 \times 5 + 7 \times 3$, in preparation for learning about the distributive property. In the expression $x^2 + 9x + 14$, older students can see the 14 as 2×7 and the 9 as $2 + 7$. They recognize the significance of an existing line in a geometric figure and can use the strategy of drawing an auxiliary line for solving problems. They also can step back for an overview and shift perspective. They can see complicated things, such as some algebraic expressions, as single objects or as being composed of several objects. For example, they can see $5 - 3(x - y)^2$ as 5 minus a positive number times a square and use that to realize that its value cannot be more than 5 for any real numbers x and y.	**GMP 7.1** Find, extend, analyze, and create patterns. **GMP 7.2** Use patterns and structures to solve problems.
Standard for Mathematical Practice 8: Look for and express regularity in repeated reasoning.	
Mathematically proficient students notice if calculations are repeated, and look both for general methods and for shortcuts. Upper elementary students might notice when dividing 25 by 11 that they are repeating the same calculations over and over again, and conclude they have a repeating decimal. By paying attention to the calculation of slope as they repeatedly check whether points are on the line through (1, 2) with slope 3, middle school students might abstract the equation $(y - 2)/(x - 1) = 3$. Noticing the regularity in the way terms cancel when expanding $(x - 1)(x + 1)$, $(x - 1)(x^2 + x + 1)$, and $(x - 1)(x^3 + x^2 + x + 1)$ might lead them to the general formula for the sum of a geometric series. As they work to solve a problem, mathematically proficient students maintain oversight of the process, while attending to the details. They continually evaluate the reasonableness of their intermediate results.	**GMP 8.1** Use patterns and structures to create and explain rules and shortcuts. **GMP 8.2** Use properties, rules, and shortcuts to solve problems. **GMP 8.3** Reflect on your thinking before, during, and after you solve a problem.

Integrating the Standards for Mathematical Practice into Classroom Instruction

After writing the Goals for Mathematical Practice, the authors next worked to clarify how teachers could use the rich activities in *Everyday Mathematics* to integrate CCSS's ambitious Standards for Mathematical Practice into their everyday teaching.

Consider, for example, a Part 3 activity in Grade 4, Lesson 2-4: Deciphering a Place-Value Code (*Teacher's Lesson Guide*, page 105).

ENRICHMENT

▶ **Deciphering a Place-Value Code**
(*Math Masters*, p. 50)

PARTNER ACTIVITY
15–30 Min

To apply students' understanding of place value, have them decipher the packing-system code used at a bakery. Encourage students to use a visual organizer such as the following to help them solve the problem. Students might begin by asking "How many boxes of *x* muffins?" beginning with 27, then 9, and so on.

Total Muffins	Boxes of 27	Boxes of 9	Boxes of 3	Boxes of 1

Have students describe how the chart they used to solve the problem is different from and similar to a base-ten place-value chart. In this problem, you multiply by 3 to get the next column. In the base-ten place-value chart, you multiply by 10. In both charts, each time you have enough in one column, that column becomes 0 and the next column becomes 1.

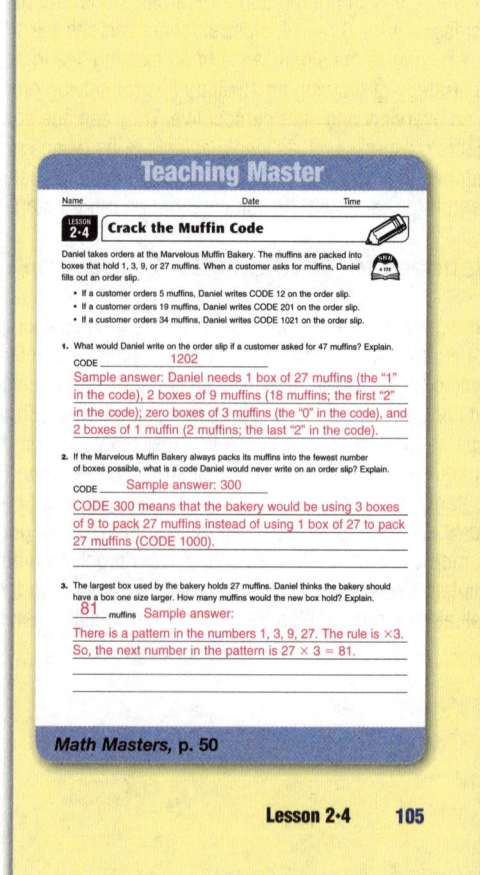

Teaching Master

Name   Date   Time

LESSON 2·4 Crack the Muffin Code

Daniel takes orders at the Marvelous Muffin Bakery. The muffins are packed into boxes that hold 1, 3, 9, or 27 muffins. When a customer asks for muffins, Daniel fills out an order slip.

- If a customer orders 5 muffins, Daniel writes CODE 12 on the order slip.
- If a customer orders 19 muffins, Daniel writes CODE 201 on the order slip.
- If a customer orders 34 muffins, Daniel writes CODE 1021 on the order slip.

1. What would Daniel write on the order slip if a customer asked for 47 muffins? Explain.
   CODE __1202__
   Sample answer: Daniel needs 1 box of 27 muffins (the "1" in the code), 2 boxes of 9 muffins (18 muffins; the first "2" in the code); zero boxes of 3 muffins (the "0" in the code), and 2 boxes of 1 muffin (2 muffins; the last "2" in the code).

2. If the Marvelous Muffin Bakery always packs its muffins into the fewest number of boxes possible, what is a code Daniel would never write on an order slip? Explain.
   CODE __Sample answer: 300__
   CODE 300 means that the bakery would be using 3 boxes of 9 to pack 27 muffins instead of using 1 box of 27 to pack 27 muffins (CODE 1000).

3. The largest box used by the bakery holds 27 muffins. Daniel thinks the bakery should have a box one size larger. How many muffins would the new box hold? Explain.
   __81__ muffins Sample answer:
   There is a pattern in the numbers 1, 3, 9, 27. The rule is ×3. So, the next number in the pattern is 27 × 3 = 81.

Math Masters, p. 50

Lesson 2·4   105

This activity connects to eight *Everyday Mathematics* GMPs, which correspond to five different CCSS-M Standards for Mathematical Practice. Below is a list of these goals and a brief explanation of how they are addressed within the activity.

Everyday Mathematics Goal for Mathematical Practice	Rationale
GMP 1.2 Make a plan for solving your problem.	Students plan how they will decipher the code.
GMP 1.3 Try different approaches when your problem is hard.	Students try different ways of deciphering the code if their first method doesn't work.
GMP 3.2 Work to make sense of others' mathematical thinking.	Students listen to others explain how they solved the problem and how the problem compares to a base-ten place-value system.
GMP 4.1 Apply mathematical ideas to real-world situations.	Students identify a pattern in order to figure out the packing-system code used at a bakery.
GMP 4.2 Use mathematical models such as graphs, drawings, tables, symbols, numbers, and diagrams to solve problems.	Students are encouraged to use a visual organizer to help them solve the problem.
GMP 6.1 Communicate your mathematical ideas clearly and precisely.	Students explain their thinking on *Math Masters*, page 50, and after they solve the problem.
GMP 7.1 Find, extend, analyze, and create patterns.	Students find and extend a pattern in order to decipher and explain the muffin code.
GMP 7.2 Use patterns and structures to solve problems.	After figuring out the pattern, students apply the pattern to solve problems on *Math Masters*, page 50.

While the table above illustrates how rich many *Everyday Mathematics* activities are, trying to address eight *Everyday Mathematics* GMPs and five Standards for Mathematical Practice in a single activity would be overwhelming. For this reason, the *Everyday Mathematics* authors decided that a focused approach would be a more productive one that would provide specific guidance about how to address one GMP at a time. With such an approach, over time, teachers would gain facility in promoting student growth in all the Standards for Mathematical Practice and would be able to take advantage of opportunities as they arise in the curriculum.

Opportunities in *Everyday Mathematics®* for Addressing the Goals for Mathematical Practice

In order to better focus attention on particular Standards for Mathematical Practice, the *Everyday Mathematics* authors created charts that show how selected lesson activities in the first two units of every grade (K–6) can be used to address specific *Everyday Mathematics* Goals for Mathematical Practice (GMPs). The activities come from Parts 1 and 2 of the lessons and are only a sampling of possible opportunities in which the Goals for Mathematical Practice could be addressed. Although the optional Part 3 activities are not included, they are a rich resource for addressing the GMPs. Therefore, teachers are highly encouraged to draw from Part 3 activities when thinking about these practice goals.

The charts are organized into four columns:

◆ The first column lists the title of the activity and the *Teacher's Lesson Guide* pages where the activity can be found.

◆ The second column highlights one GMP that is addressed in the activity. Many activities address more than one GMP. The goal identified in the second column was selected because it is clearly and strongly addressed in that particular activity. Highlighting only one goal also allows teachers to see how they might focus on that specific goal in depth. Additional practice goals for this activity (marked "*See also*") are listed in this column. Teachers may choose to focus their instruction on any of these goals.

◆ The third column of the chart provides a brief description of the part of the activity that provides an opportunity to address the GMP highlighted in the second column.

◆ The final column of the chart, Guiding Questions, is designed to help teachers raise the level of mathematical discourse in their classrooms. There are two types of Guiding Questions: those that are specific to the activity (in regular type) and those that are global, overarching questions (in boldface type). The specific questions provide support for teachers as they use the content of the activity to focus instruction on the GMP highlighted in the second column. The questions listed in boldface are not specific to the activity itself, but are broad questions that can be generalized to many different mathematical activities and situations. Initially, students may struggle to find an appropriate response to these broad questions. However, there are various entry points to these questions that can lead to a rich discussion. A teacher might model appropriate responses to these questions and identify how the activity students completed relates to the broad question. These questions are meant to be revisited over time. As a result, students will continue to develop proficiencies with the mathematical practices, and their answers to these questions should become deeper and richer.

The Guiding Questions can be used in a variety of ways: They can be presented to students before an activity; asked during the activity; used to stimulate a summary discussion; posed as a prompt for an Exit Slip; and other ways not listed here. Note that there is no "right" way to use these questions. The way that they are used should be left to the teacher's best judgment. The Guiding Questions listed in the chart are also merely illustrative. Teachers should feel free to substitute other questions as they deem appropriate. More important than choosing the specific question is establishing norms supportive of the mathematical practices at the beginning of the school year. With the establishment of such norms early in the year, the type of questioning suggested in the chart will become an integral part of classroom instruction all year long.

Everyday Mathematics® Goals for Mathematical Practice

Practice Standard 1: Make sense of problems and persevere in solving them.

GMP 1.1 Work to make sense of your problem.
GMP 1.2 Make a plan for solving your problem.
GMP 1.3 Try different approaches when your problem is hard.
GMP 1.4 Solve your problem in more than one way.
GMP 1.5 Check whether your solution makes sense.
GMP 1.6 Connect mathematical ideas and representations to one another.

Practice Standard 2: Reason abstractly and quantitatively.

GMP 2.1 Represent problems and situations mathematically with numbers, words, pictures, symbols, gestures, tables, graphs, and concrete objects.
GMP 2.2 Explain the meanings of the numbers, words, pictures, symbols, gestures, tables, graphs, and concrete objects you and others use.

Practice Standard 3: Construct viable arguments and critique the reasoning of others.

GMP 3.1 Explain both what to do and why it works.
GMP 3.2 Work to make sense of others' mathematical thinking.

Practice Standard 4: Model with mathematics.

GMP 4.1 Apply mathematical ideas to real-world situations.
GMP 4.2 Use mathematical models such as graphs, drawings, tables, symbols, numbers, and diagrams to solve problems.

Practice Standard 5: Use appropriate tools strategically.

GMP 5.1 Choose appropriate tools for your problem.
GMP 5.2 Use mathematical tools correctly and efficiently.
GMP 5.3 Estimate and use what you know to check the answers you find using tools.

Practice Standard 6: Attend to precision.

GMP 6.1 Communicate your mathematical thinking clearly and precisely.
GMP 6.2 Use the level of precision you need for your problem.
GMP 6.3 Be accurate when you count, measure, and calculate.

Practice Standard 7: Look for and make use of structure.

GMP 7.1 Find, extend, analyze, and create patterns.
GMP 7.2 Use patterns and structures to solve problems.

Practice Standard 8: Look for and express regularity in repeated reasoning.

GMP 8.1 Use patterns and structures to create and explain rules and shortcuts.
GMP 8.2 Use properties, rules, and shortcuts to solve problems.
GMP 8.3 Reflect on your thinking before, during, and after you solve a problem.

Kindergarten Section 1

Activity	*Everyday Mathematics* Goal for Mathematical Practice	Opportunity	Guiding Questions
Section 1			
Activity 1♦1: Partner March			
Matching Strips (*Teacher's Guide to Activities*, pages 46 and 47)	**GMP 3.1** Explain both what to do and why it works. See also: **GMP 6.1**	Children use strips of paper to find lengths that are the same, longer, or shorter.	How did you find out if your strips were the same length? How do you know if something is longer or shorter than something else? **Why (or when) might we need to know how to compare lengths of objects?**
Activity 1♦2: Introduction to Pattern Blocks			
Exploring Pattern Blocks (*Teacher's Guide to Activities*, pages 48 and 49)	**GMP 6.1** Communicate your mathematical thinking clearly and precisely. See also: **GMP 7.1**	Children explore the shapes of pattern blocks in open-ended ways.	How many sides does your shape have? Can you find another shape with the same number of sides? Do all the shapes with the same number of sides look the same? How are they the same or different? **How could you describe your shape to someone who can't see it or feel it?**
Activity 1♦3: Multisensory Counts			
Counting by Touch and Sound (*Teacher's Guide to Activities*, pages 50 and 51)	**GMP 2.1** Represent problems and situations mathematically with numbers, words, pictures, symbols, gestures, tables, graphs, and concrete objects. See also: **GMP 1.6, GMP 6.3**	Children practice counting with sensory activities.	Are 4 objects (or another number) the same as 4 taps or 4 sounds? What makes them all "4"? How else could we show "4" (or another number)? **How do you make sure you count correctly whatever you are counting?**
Activity 1♦4: Countdown to Zero			
Singing and Eating Down to Zero (*Teacher's Guide to Activities*, pages 52 and 53)	**GMP 2.2** Explain the meaning of the numbers, words, pictures, symbols, gestures, tables, graphs, and concrete objects you and others use. See also: **GMP 4.1**	Children explore the meaning of the number zero through songs and snacks.	Where else have you seen the number zero before? **Why do you think we have the number zero?**

Kindergarten Section 1 (cont.)

Activity	*Everyday Mathematics* Goal for Mathematical Practice	Opportunity	Guiding Questions
Activity 1◆5: Getting to Know Numbers (1–9)			
Exploring Featured Numbers (*Teacher's Guide to Activities,* pages 54, 55, and 55A)	**GMP 1.6** Connect mathematical ideas and representations to one another. *See also:* **GMP 2.1, GMP 2.2, GMP 6.3, GMP 7.1**	Children explore and represent a featured number in a variety of ways.	How can all the creations be "4" (or a different number) but look so different? Do you notice anything similar about the creations for 4 at our number stations and on our number poster? **What makes all these representations "4"?** **How do you make sure you count correctly whatever you are counting?**
Activity 1◆6: Introduction to Sorting			
Sorting by Attributes (*Teacher's Guide to Activities,* pages 56 and 57)	**GMP 7.1** Find, extend, analyze, and create patterns. *See also:* **GMP 6.3**	Children sort collections of objects by attributes.	What are some ways we can sort the objects? How many different ways can we sort the objects? **How do you decide a good way to sort?**
Activity 1◆7: Sand and Water Play			
Experimenting with Volume (*Teacher's Guide to Activities,* pages 58 and 59)	**GMP 3.1** Explain both what to do and why it works. *See also:* **GMP 1.2, GMP 4.1**	Children explore ways to compare the volumes of different containers and discuss their thinking.	Which container do you think holds more? Why do you think so? How can we find out which container holds more? Did our plan work? **How else might we compare the volumes of the containers?**

Kindergarten Section 1 (cont.)

Activity	*Everyday Mathematics* Goal for Mathematical Practice	Opportunity	Guiding Questions
Activity 1♦8: Birthday Graphs			
Graphing Birthdays and Ages (*Teacher's Guide to Activities,* pages 60 and 61)	**GMP 4.2** Use mathematical models such as graphs, drawings, tables, symbols, numbers, and diagrams to solve problems. See also: GMP 2.1, GMP 4.1	Children create a class birthday graph and use it to answer questions.	Can we learn something from the graph without counting? What questions can we answer by counting the cakes in the graph? **How does a graph help us answer questions?**
Activity 1♦9: Sound and Motion Patterns			
Discovering Patterns (*Teacher's Guide to Activities,* pages 62 and 63)	**GMP 7.1** Find, extend, analyze, and create patterns.	Children replicate and extend patterns using sounds and motions.	What should we add next to continue the pattern? How do you know? **How do you know if something is a pattern?** **What else can we use to make patterns besides sounds and movements?**
Activity 1♦10: Patterns with Color			
Creating and Extending Patterns (*Teacher's Guide to Activities,* pages 64 and 65)	**GMP 3.1** Explain both what to do and why it works. See also: GMP 7.1	Children create patterns using colored objects and discuss characteristics of patterns.	What do you notice about the pattern? What should we add next to continue the pattern? How do you know? **How do you know if something is a pattern?** **How can you describe this pattern?**
Activity 1♦11: Coin Comparisons			
Sorting Coins into "Banks" (*Teacher's Guide to Activities,* pages 66 and 67)	**GMP 3.2** Work to make sense of others' mathematical thinking. See also: GMP 3.1, GMP 6.3, GMP 7.1	Children sort coins and share their groupings.	Is there more than one way to sort the coins? Look at someone else's sorting. Can you figure out how they sorted? Was it the same or different than your way? **What can you do if you don't understand how someone else sorted the coins?**

Kindergarten Section 1 (cont.)

Activity	*Everyday Mathematics* Goal for Mathematical Practice	Opportunity	Guiding Questions
Activity 1♦12: *Give the Next Number* Game			
Playing *Give the Next Number* (*Teacher's Guide to Activities*, pages 68 and 69)	**GMP 6.3** Be accurate when you count, measure, and calculate. *See also:* **GMP 2.1**	Children play a game to practice counting and counting on.	How do you keep track when it's not your turn to say the number? **Does it matter in what order we say the numbers when we count? Why or why not?** **In real life, why is it important to count correctly?**
Activity 1♦13: Body Height Comparisons			
Comparing Body Heights to Objects (*Teacher's Guide to Activities*, pages 70 and 71)	**GMP 4.1** Apply mathematical ideas to real-world situations. *See also:* **GMP 2.1, GMP 4.2**	Children compare their heights to objects in the classroom and display their findings.	How does our display show information about the heights of things in our classroom? **Why might we want to know which things in our classroom are taller or shorter than we are? How could we use this information?** **When else might we want to know which things are taller or shorter than we are?**
Activity 1♦14: Finger Count Fun			
Reviewing Numbers (*Teacher's Guide to Activities*, pages 72 and 73)	**GMP 2.1** Represent problems and situations mathematically with numbers, words, pictures, symbols, gestures, tables, graphs, and concrete objects. *See also:* **GMP 1.6, GMP 6.3, GMP 7.1**	Children discuss how numbers are represented in a counting book.	How did the author of this book show (or represent) each number? Can you think of other ways to show or act out the number on this page? **If you made a counting book, how might you represent the numbers?**

Kindergarten Section 1 (cont.)

Activity	*Everyday Mathematics* Goal for Mathematical Practice	Opportunity	Guiding Questions
Activity 1◆15: Shape Puzzles			
Combining and Creating Shapes (*Teacher's Guide to Activities*, pages 73A and 73B)	**GMP 1.4** Solve your problem in more than one way. See also: **GMP 3.2, GMP 6.1**	Children combine shapes in many ways to make other shapes.	How many different shapes can you make using rectangles (or another shape)? How many ways did you find to make a rectangle using only triangles? **What do you do to find different ways to make the same shape?** **Did you notice any classmates combining shapes in different ways than you did? What did they do?**
Activity 1◆16: Ten Frames			
Exploring Ten Frames (*Teacher's Guide to Activities*, pages 73C and 73D)	**GMP 2.2** Explain the meanings of the numbers, words, pictures, symbols, gestures, tables, graphs, and concrete objects you and others use. See also: **GMP 2.1, GMP 3.1, GMP 5.2, GMP 6.3, GMP 7.2**	Children explore ten frames and discuss the way numbers are represented on them.	What do you notice about a ten frame? Why aren't there any counters in the bottom row for the numbers less than 5? What do you notice about the ten frames for two numbers in a row, like 1 and 2, or 7 and 8? Why do you think that is? **How can ten frames help us show or "see" numbers?**

Kindergarten Section 2

Activity	*Everyday Mathematics* Goal for Mathematical Practice	Opportunity	Guiding Questions
Activity 2◆1: Shape Collages			
Exploring Shapes (*Teacher's Guide to Activities,* pages 88 and 89)	**GMP 4.1** Apply mathematical ideas to real-world situations. *See also:* **GMP 2.1, GMP 6.1, GMP 7.1**	Children look for shapes in pictures and objects.	What other objects can you think of that have this shape? Why do you think an analog clock (or other object) is a circle (or other shape)? What if it were a different shape? **Why might the shape of the object matter?**
Activity 2◆2: Shapes by Feel			
Identifying Attributes of Shapes (*Teacher's Guide to Activities,* pages 90 and 91)	**GMP 3.2** Work to make sense of others' mathematical thinking. *See also:* **GMP 3.1, GMP 4.1, GMP 6.1, GMP 7.1**	Children identify shapes by touch and verbal descriptions.	What would someone need to tell you about a shape so that you could name it without seeing it? Let's have the shape feeler give us clues about how the shape feels so we can try to guess the shape without looking: Can you tell what shape it is? Why did you guess that shape? Could it have been any other shape? **What might you do if you don't understand someone's description of the block or if you need more information? What questions might you ask?** **Why is it important to try to make sense of what others are describing?**
Activity 2◆3: Which Way Do I Go?			
Completing an Obstacle Course (*Teacher's Guide to Activities,* pages 92 and 93)	**GMP 6.1** Communicate your mathematical thinking clearly and precisely. *See also:* **GMP 4.1, GMP 6.2**	Children describe obstacle courses for others to follow.	What words help you tell what to do for the obstacle course? **When else do you use position words? Why are position words useful?** **What happens if you don't give clear directions?**

Kindergarten Section 2 (cont.)

Activity	*Everyday Mathematics* Goal for Mathematical Practice	Opportunity	Guiding Questions
Activity 2♦4: *Spin a Number* Game			
Making and Playing *Spin a Number* (*Teacher's Guide to Activities,* pages 94 and 95)	**GMP 6.3** Be accurate when you count, measure, and calculate.	Children count a given number of spaces on a gameboard.	How do you know how many spaces to move on the gameboard? How do you keep track of the number of spaces you have moved? **Why is it important to count carefully (accurately)?**
Activity 2♦5: Patterns All Around			
Looking for Patterns (*Teacher's Guide to Activities,* pages 96 and 97)	**GMP 7.1** Find, extend, analyze, and create patterns. *See also:* **GMP 4.1**	Children find and describe patterns in pictures and their environment.	Describe your pattern. Look at where the pattern stops. What would come next? How do you know? **How do you know when you've found a pattern?** **What is a pattern?**
Activity 2♦6: Playful Oral Counting Games			
Playing Counting Games (*Teacher's Guide to Activities,* pages 98 and 99)	**GMP 5.1** Choose appropriate tools for your problem. *See also:* **GMP 6.3**	Children play counting games and discuss the usefulness of tools, such as their fingers, a number line, or a number grid, to help them count.	If you are unsure, what tools in our classroom could you use to help you keep track of the count? If we play using higher numbers or skip counting, what tools could we use? Why would those be good choices? **What are some other useful counting tools? How are they helpful?** **When do you need a tool to help you count? When do you not need a tool?**
Activity 2♦7: Preparation for Number Writing			
Getting to Know Numbers, 1–9 (*Teacher's Guide to Activities,* page 103)	**GMP 6.2** Use the level of precision you need for your problem. *See also:* **GMP 2.1, GMP 3.1, GMP 6.2, GMP 6.3**	Children compare and order sets of objects from smallest to largest.	Compare your set of objects to someone else's. Who has more? How did you figure it out? How can you figure out how many more or less you have? How can we put the sets in order from smallest to largest? **Do you always need to count to compare or order sets? Why or why not?**

Kindergarten Section 2 (cont.)

Activity	*Everyday Mathematics* Goal for Mathematical Practice	Opportunity	Guiding Questions
Activity 2◆8: *Matching Coin Game*			
Playing the *Matching Coin Game* (*Teacher's Guide to Activities,* pages 104 and 105)	**GMP 6.3** Be accurate when you count, measure, and calculate. See also: **GMP 4.1**	Children compare collections of coins.	How do you know which set has the most coins? How many more dimes (or other coins) do you have than pennies (or other coins)? How did you figure it out? **What do you do to help you count correctly (accurately)?** **Why is it important to count accurately?**
Activity 2◆9: Number Board			
Building a Number Board (*Teacher's Guide to Activities,* pages 106 and 107)	**GMP 7.1** Find, extend, analyze, and create patterns. See also: **GMP 2.1, GMP 2.2, GMP 6.1, GMP 6.3**	Children describe a number board and discuss how the digits 0–9 can be used to write any counting number.	What do you notice about the number board? Do you see any patterns? If we added numbers to the board, what would the dots look like? What digits would you see? **What other patterns have you noticed in numbers?**
Activity 2◆10: Tricky Teens			
Introducing the *Tricky Teens Game* (*Teacher's Guide to Activities,* pages 108 and 109)	**GMP 7.2** Use patterns and structures to solve problems. See also: **GMP 2.2**	Children look for patterns in the teen numbers to put them in order from 10 to 19.	What is the same about these numbers (10–19)? How are the teen numbers different from the numbers 1 through 9? How are they similar to the numbers 1–9? **Do you notice any patterns in the numbers 0 through 19?** **How do these patterns help us put the teen numbers in order?**
Activity 2◆11: Listen and Do (10–19)			
Counting and Moving (*Teacher's Guide to Activities,* pages 110 and 111)	**GMP 2.2** Explain the meaning of the numbers, words, pictures, symbols, gestures, tables, graphs, and concrete objects you and others use. See also: **GMP 2.1, GMP 6.3**	Children act out numbers with physical actions.	How do your claps (or other movements) show (represent) the number on your card? How does the number on your card represent your claps and movements? **What does the number on your card mean?** **Why is it useful to be able to represent sounds, actions, and other things with numbers?**

Kindergarten Section 2 (cont.)

Activity	*Everyday Mathematics* Goal for Mathematical Practice	Opportunity	Guiding Questions
Activity 2◆12: Teen Partners			
Representing Teen Numbers (*Teacher's Guide to Activities,* pages 112 and 113)	**GMP 8.1** Use patterns and structures to create and explain rules and shortcuts. *See also:* GMP 1.6, GMP 2.1, GMP 2.2, GMP 3.1, GMP 7.2	Pairs of children represent teen numbers using their fingers and discuss how these numbers can be thought of as "10 and some more."	Can we show all of the teen numbers using two children's hands? Why or why not? If you always use ten fingers on one child's hands, how do you know how many fingers the other child should show? **Why or when might it be helpful to think of teen numbers as "10 and some more?" Are there other numbers you can think of in similar ways?**
Activity 2◆13: Estimation Jars			
Making an Estimate (*Teacher's Guide to Activities,* pages 114 and 115)	**GMP 1.5** Check whether your solution makes sense. *See also:* GMP 1.2, GMP 2.1, GMP 3.1, GMP 5.2, GMP 6.2	Children estimate the number of objects in a jar using a jar of 10 of the same objects as a reference.	Why do you think your estimate is a good one? What can you do to check if your estimate makes sense? How can the jar of 10 objects help you check your estimate? **Why should you check if your estimate makes sense?** **How can you get better at making estimates?**
Activity 2◆14: Number Stories: Stage 1			
Telling and Acting Out Number Stories (*Teacher's Guide to Activities,* pages 116–118)	**GMP 1.1** Work to make sense of your problem. *See also:* GMP 1.2, GMP 1.4, GMP 1.6, GMP 2.1, GMP 2.2, GMP 3.1, GMP 3.2, GMP 4.1, GMP 5.2, GMP 6.1	Children solve number stories in various ways.	How did you make sense of this number story? **Why is it important to understand a number story before you solve it? What helps you understand a number story?** **What can you do if you don't understand a number story?**

Kindergarten Section 2 (cont.)

Activity	*Everyday Mathematics* Goal for Mathematical Practice	Opportunity	Guiding Questions
Activity 2♦15: Symmetry Painting			
Combining and Creating Shapes (*Teacher's Guide to Activities,* page 121)	**GMP 1.3** Try different approaches when your problem is hard. See also: **GMP 1.1, GMP 1.2, GMP 1.4, GMP 3.1, GMP 6.1**	Children find many ways to combine shapes to make new shapes.	How do you discover new shapes to make with the same cards? What can you do if you're having trouble finding another way to combine your shapes? **Why is it good to stick with a problem, even if it's hard? How do you help yourself do that?**
Activity 2♦16: Symmetry in Nature			
Creating a Bar Graph (*Teacher's Guide to Activities,* page 123)	**GMP 4.2** Use mathematical models such as graphs, drawings, tables, symbols, numbers, and diagrams to solve problems. See also: **GMP 2.1, GMP 2.2, GMP 4.1, GMP 6.1, GMP 6.3**	Children create a bar graph and use it to answer questions about their favorite school activities.	Can you think of a specific question or problem we might ask or solve using the graph? How can the rest of us solve that problem or question? What things can we learn from the graph without counting the totals for each activity? What can we find out by counting the totals for each activity? **Why did we make a graph to learn about our favorite school activities? When else might we make graphs?**

Grade 1 Unit 1
Establishing Routines

Activity	*Everyday Mathematics* Goal for Mathematical Practice	Opportunity	Guiding Questions
Lesson 1♦1: Daily Routines			
Counting the Days of School (*Teacher's Lesson Guide,* pages 17 and 18)	GMP 5.2 Use mathematical tools correctly and efficiently. See also: GMP 1.6, GMP 2.1, GMP 4.1	Children learn to use the number line to count the days in school.	How can you use the number line to help find the number of days we've been in school? Describe another tool and how you could use it to count the days in school. **What other ways can you use the number line?** **Why is it helpful to use a tool for counting?**
Lesson 1♦2: Investigating the Number Line			
Finding the Number of Children Who Are Absent (*Teacher's Lesson Guide,* pages 22 and 23)	GMP 3.2 Work to make sense of others' mathematical thinking. See also: GMP 1.5, GMP 3.1, GMP 5.2	Children share strategies to find how many children are absent today.	Tell about a classmate's strategy for solving the problem that is different from your own. **Why is it important to understand how another person solved a problem?** **What can you do if you don't understand how someone else solved the problem?**
Lesson 1♦3: Tools for Doing Mathematics			
Using the Pattern-Block Template (*Teacher's Lesson Guide,* page 26)	GMP 5.2 Use mathematical tools correctly and efficiently.	Children explore the Pattern-Block Template and practice using it.	What kinds of things could you do using the Pattern-Block Template? How do you think you might use the Pattern-Block Template correctly? **When might you use a tool to solve a problem?**
Lesson 1♦4: Number-Writing Practice			
Writing the Numbers 1 and 2 (*Teacher's Lesson Guide,* page 31)	GMP 6.1 Communicate your mathematical ideas clearly and precisely. See also: GMP 1.6, GMP 2.1	Children practice writing the numerals 1 and 2.	Why do you need to be able to write numbers? How should you write a 1 so that you and others can read it? How should you write a 2 so that you and others can read it? **Why is it important that you can read the numbers you write?** **Why is it important that others can read the numbers you write?**

Grade 1 Unit 1
Establishing Routines (cont.)

Activity	*Everyday Mathematics* Goal for Mathematical Practice	Opportunity	Guiding Questions
Lesson 1♦5: One More, One Less			
Telling "One More" and "One Less" Stories (*Teacher's Lesson Guide*, pages 34 and 35)	**GMP 1.1** Work to make sense of your problem. *See also:* **GMP 4.1, GMP 5.2, GMP 6.1**	Children solve problems using the number line.	How do you know whether to move forward or backward on the number line to solve each problem? **What information in the problem is important?** **What can you do if you don't understand a problem?**
Lesson 1♦6: Comparing Numbers			
Comparing and Ordering Numbers (*Teacher's Lesson Guide*, page 38)	**GMP 2.2** Explain the meaning of the numbers, words, pictures, symbols, gestures, tables, graphs, and concrete objects you and others use.	Children compare and order numbers.	Explain how you know which is the larger number. **How could you explain to a friend the meaning of the number 12 (or another number)?**
Lesson 1♦7: Recording Tally Counts			
Making a Tally Chart to Count Children's Pets (*Teacher's Lesson Guide*, page 44)	**GMP 4.2** Use mathematical models such as graphs, drawings, tables, symbols, numbers, and diagrams to solve problems. *See also:* **GMP 2.1, GMP 4.1**	Children use a tally chart to represent data they collect to find out how many different pets are owned by everyone in the class.	What does each tally in the chart represent? What questions can you answer using the information in this tally chart? **For what other purpose could you use a tally chart?** **Why is it useful to put the information (data) in a tally chart?**
Lesson 1♦8: Investigating Equally Likely Outcomes			
Dice-Roll and Tally (*Teacher's Lesson Guide*, pages 48 and 49)	**GMP 8.1** Use patterns and structures to create and explain rules and shortcuts. *See also:* **GMP 2.1, GMP 2.2, GMP 3.2, GMP 4.1, GMP 4.2**	Children discuss the patterns in a cumulative tally of the data they collect.	What do you notice about the total number of times each number was rolled? **Is there a pattern in the number of times each number was rolled? Describe the pattern.** **Use the pattern to create a rule about which number will come up more often when you roll a die. Describe your rule.**

Grade 1 Unit 1
Establishing Routines (cont.)

Activity	*Everyday Mathematics* Goal for Mathematical Practice	Opportunity	Guiding Questions
Lesson 1♦9: The Calendar			
Introducing the Class Calendar (*Teacher's Lesson Guide,* pages 53 and 54)	**GMP 4.1** Apply mathematical ideas to real-world situations. *See also:* **GMP 5.2**	Children discuss the uses of calendars and the elements of a monthly calendar.	When have you seen someone use a calendar? Describe how they used it. What can we find out by looking at this month's calendar? **Now that you know how to use a calendar, how could it be helpful in your everyday life?**
Lesson 1♦10: Working in Small Groups			
Playing *Top-It* in Small Groups (*Teacher's Lesson Guide,* pages 57 and 58)	**GMP 2.2** Explain the meaning of the numbers, words, pictures, symbols, gestures, tables, graphs, and concrete objects you and others use.	Children compare numbers to decide which is largest.	How do you decide which number is largest? What can you do if you don't know which number is largest? **What do you picture in your mind when you think about the number 13 (or another number)? How can you use this picture to help you compare numbers?**
Lesson 1♦11: Explorations: Exploring Math Materials			
Exploring with Pattern Blocks, Base-10 Blocks, and Geoboards (*Teachers Lesson Guide,* pages 61 and 62)	**GMP 5.2** Use mathematical tools correctly and efficiently.	Children explore pattern blocks, base-10 blocks, and geoboards as tools for mathematics.	How can you use pattern blocks as tools to do math? base-10 blocks? geoboards? How do you think you might use pattern blocks correctly? base-10 blocks? geoboards? **When might you use a tool to solve a problem?**

Grade 1 Unit 1
Establishing Routines (cont.)

Activity	*Everyday Mathematics* Goal for Mathematical Practice	Opportunity	Guiding Questions
Lesson 1♦12: Weather and Temperature Routines			
Introducing the Daily Temperature Routine (*Teachers Lesson Guide,* pages 67 and 68)	**GMP 6.2** Use the level of precision you need for your problem. *See also:* **GMP 4.1, GMP 5.2, GMP 6.3**	Children use appropriately precise language to report the temperature to the nearest 10 degrees.	What can you tell about the temperature when reading the thermometer to the nearest 10 degrees? **In what situations is it important to tell an exact temperature? When is an estimate okay?**
Lesson 1♦13: Number Stories			
Telling Simple Number Stories (*Teachers Lesson Guide,* pages 72 and 73)	**GMP 1.5** Check whether your solution makes sense. *See also:* **GMP 1.1, GMP 1.2, GMP 1.3, GMP 1.4, GMP 2.1, GMP 2.2, GMP 3.1, GMP 3.2, GMP 4.1, GMP 4.2**	Children solve number stories and share answers and solution strategies.	How can you check whether your solution makes sense? What can you do if your answer doesn't make sense? **If your answer is different from someone else's, how can you determine which answer is correct?**

Grade 1 Unit 2
Everyday Uses of Numbers

Activity	*Everyday Mathematics* Goal for Mathematical Practice	Opportunity	Guiding Questions
Lesson 2♦1: Number Grids			
Playing *Rolling for 50* (*Teacher's Lesson Guide,* page 95)	**GMP 1.6** Connect mathematical ideas and representations to one another. See also: **GMP 5.2**	Children compare the number grid and the number line before playing *Rolling for 50.*	What do you notice about the number grid? How are the number line and the number grid the same? How are they different? **Why it is important to be able to see counting in different ways?**
Lesson 2♦2: Numbers All Around			
Reviewing Facts Within 5 (*Teacher's Lesson Guide,* pages 101 and 102)	**GMP 8.2** Use properties, rules, and shortcuts to solve problems. See also: **GMP 3.1, GMP 3.2, GMP 6.1, GMP 8.1**	Children use strategies to solve simple addition and subtraction facts.	What strategy did you use to add (or subtract) 0? What strategy did you use to add (or subtract) 1? When might you use the counting-on strategy? **Why are strategies (shortcuts) helpful for solving problems?**
Lesson 2♦3: Complements of 10			
Introducing Ten Frames (*Teacher's Lesson Guide,* page 106)	**GMP 2.1** Represent problems and situations mathematically with numbers, words, pictures, symbols, gestures, tables, graphs, and concrete objects. See also: **GMP 2.2, GMP 5.2**	Children use a ten frame to represent numbers.	What can you tell about the number 5 (or another number) by looking at the ten frame? What do you notice about all of the numbers larger than 5? **What can you learn about numbers when you show them on a ten frame? Why do you think we use a ten frame instead of a frame with a different number of spaces?**
Lesson 2♦4: Unit Labels for Numbers			
Labeling Numbers with Units (*Teacher's Lesson Guide,* page 110)	**GMP 6.1** Communicate your mathematical ideas clearly and precisely. See also: **GMP 2.2, GMP 4.1**	Children discuss the importance of including labels with numbers.	When should you label a number? How do you choose a label for a number? **Why is it important to label the numbers you use? What might happen if you don't label a number? Why is it important for others to understand your mathematical ideas?**

Grade 1 Unit 2
Everyday Uses of Numbers (cont.)

Activity	*Everyday Mathematics* Goal for Mathematical Practice	Opportunity	Guiding Questions
Lesson 2◆5: Analog Clocks			
Estimating the Time Shown on an Hour-Hand-Only Clock (*Teacher's Lesson Guide,* pages 116 and 117)	**GMP 6.2** Use the level of precision you need for your problem. *See also:* **GMP 4.1, GMP 5.2, GMP 6.1**	Children discuss time as an estimate and use appropriate language to describe the time on an hour-hand-only clock.	When is it okay to tell the time to the closest hour? When might you need to tell the time to the closest minute? To the closest second? How do you decide what words to use to tell the time? **Why is it important to describe the time clearly (precisely)?**
Lesson 2◆6: Telling Time to the Hour			
Telling Time to the Nearest Hour (*Teacher's Lesson Guide,* pages 121 and 122)	**GMP 4.1** Apply mathematical ideas to real-world situations. *See also:* **GMP 5.2, GMP 6.1**	Children tell time to the nearest hour.	In what situations do you see other people telling time? **How will knowing how to tell time be useful in your everyday life?**
Lesson 2◆7: Explorations: Exploring Lengths, Straightedges, and Dominoes			
Exploration A: Estimating the Relative Lengths of Objects (*Teacher's Lesson Guide,* page 126)	**GMP 5.3** Estimate and use what you know to check the answers you find using your tools. *See also:* **GMP 5.1, GMP 5.2**	Children remember the length of a ruler to estimate the lengths of various objects.	How did you help yourself remember the length of the ruler? Why is it helpful to be able to estimate length? In this activity, you estimated length before you measured. How could you check your measurement after you measure with a ruler? **Why is it important to check your measurements?**
Lesson 2◆8: Pennies			
Introducing Cents Notation (*Teacher's Lesson Guide,* page 131)	**GMP 4.1** Apply mathematical ideas to real-world situations. *See also:* **GMP 2.1, GMP 2.2**	Children discuss the uses of pennies.	Name some things that can be bought for 1 penny. Name some things that can be bought for 10 pennies. Name some things that can be bought for 100 pennies. **How can knowing what coins are worth help you in your daily life?**

Grade 1 Unit 2
Everyday Uses of Numbers (cont.)

Activity	*Everyday Mathematics* Goal for Mathematical Practice	Opportunity	Guiding Questions
Lesson 2♦9: Nickels			
Exploring Change (*Teacher's Lesson Guide*, pages 137 and 138)	**GMP 1.4** Solve your problem in more than one way. *See also:* **GMP 1.6, GMP 2.1, GMP 2.2, GMP 3.1, GMP 4.1**	Children count change using different combinations of pennies and nickels.	When might you need to pay for something using only pennies or using only nickels and pennies? **Why is it important to be able to solve a problem in more than one way? How can solving a problem in more than one way help you find the best strategy for you?**
Lesson 2♦10: Counting Pennies and Nickels			
Counting Nickels and Pennies (*Teacher's Lesson Guide*, page 141)	**GMP 7.2** Use patterns and structures to solve problems. *See also:* **GMP 1.2, GMP 1.6, GMP 6.3**	Children use a pattern to count combinations of nickels and pennies.	Why is it easier to count the nickels before the pennies? How would your counting change if you counted the pennies first? What pattern do you use to count the nickels? How does the pattern change when you begin to count the pennies? **Name another way patterns are useful in solving problems.**
Lesson 2♦11: Number Models			
Introducing Number Models (*Teacher's Lesson Guide*, page 147)	**GMP 4.2** Use mathematical models such as graphs, drawings, tables, symbols, numbers, and diagrams to solve problems. *See also:* **GMP 1.6, GMP 2.1, GMP 2.2, GMP 4.1**	Children write number models to solve problems about pennies being dropped into a container.	How did we write a number model to show the pennies I dropped in the container? How did we know what numbers and symbols to use? **How can writing a number model help you solve a problem?**

Grade 1 Unit 2
Everyday Uses of Numbers (cont.)

Activity	*Everyday Mathematics* Goal for Mathematical Practice	Opportunity	Guiding Questions
Lesson 2♦12: Subtraction Number Models			
Introducing Subtraction Number Models (*Teacher's Lesson Guide*, pages 151 and 152)	**GMP 2.2** Explain the meaning of the numbers, words, pictures, symbols, gestures, tables, graphs, and concrete objects you and others use. See also: GMP 1.6, GMP 2.1, GMP 4.1, GMP 4.2	Children explain the meanings of the numbers and symbols in a subtraction number model.	In the number model 8 − 6 = 2, what do the numbers 8, 6, and 2 mean? What do the symbols − and = mean? **Why is it important to know what the numbers and symbols in number models mean?**
Lesson 2♦13: Number Stories			
Solving Number Stories (*Teacher's Lesson Guide*, pages 157 and 158)	**GMP 3.1** Explain what to do and why it works. See also: GMP 1.1, GMP 1.2, GMP 1.3, GMP 1.4, GMP 1.5, GMP 1.6, GMP 2.1, GMP 2.2, GMP 3.2, GMP 4.1, GMP 4.2, GMP 5.1	Children share strategies for solving number stories.	Share your strategy for solving the problem. Explain why your strategy works. **Why is it important to be able to explain how you solved a math problem?**

Grade 2 Unit 1
Numbers and Routines

Activity	*Everyday Mathematics* Goal for Mathematical Practice	Opportunity	Guiding Questions
Lesson 1◆1: Math Message and Number Sequences			
Number of School Days Routine (*Teacher's Lesson Guide*, page 19)	**GMP 4.1** Apply mathematical ideas to real-world situations. *See also:* **GMP 1.6, GMP 2.1, GMP 2.2**	Children connect the number of school days to a number line and discuss other real-world applications for number lines.	When might you use a number line? **What are some tools that have markings similar to number lines?**
Lesson 1◆2: Tools and Coins			
Counting Coins (*Teacher's Lesson Guide*, page 26)	**GMP 2.2** Explain the meaning of the numbers, words, pictures, symbols, gestures, tables, graphs, and concrete objects you use. *See also:* **GMP 4.1, GMP 6.3**	Children identify the value of individual coins and add them together to find the value of coin combinations.	Explain why Q, D, N, and P are used to represent coins. What does the ¢ symbol mean? **Why is it important to be able to explain what numbers and symbols mean?**
Lesson 1◆3: Calendars and Clocks			
Telling Time (*Teacher's Lesson Guide*, pages 30 and 31)	**GMP 5.2** Use mathematical tools correctly and efficiently. *See also:* **GMP 2.1, GMP 2.2, GMP 3.1, GMP 4.1**	Children review how to tell time and use their tool-kit clocks to show different times.	Describe how you show the time on your tool-kit clock. **Why is it important to use mathematical tools correctly?**
Lesson 1◆4: Partner Study Routines			
Exploring Number-Grid Patterns (*Teacher's Lesson Guide*, page 36)	**GMP 7.1** Find, extend, analyze, and create patterns. *See also:* **GMP 5.2, GMP 6.1, GMP 7.1, GMP 7.2**	Children describe patterns they find on a number grid and use it to practice counting-up and counting-back strategies.	What are some of the patterns you see on the number grid? How do the patterns on the number grid help you count up and count back? **Where can you find patterns in mathematics?**
Lesson 1◆5: Grouping by Tens—$1, $10, $100			
Playing the *Money Exchange Game* with $100 Bills (*Teacher's Lesson Guide*, page 40)	**GMP 7.2** Use patterns and structures to solve problems. *See also:* **GMP 2.1, GMP 4.1, GMP 6.3**	Children practice counting and exchanging money by playing the *Money Exchange Game*. Children use the base-10 number system when exchanging bills.	How do you know when to make an exchange? What is the relationship between $1 bills, $10 bills, and $100 bills? **When is the relationship between ones, tens, and hundreds important in mathematics?**

Grade 2 Unit 1
Numbers and Routines (cont.)

Activity	*Everyday Mathematics* Goal for Mathematical Practice	Opportunity	Guiding Questions
Lesson 1◆6: Math Boxes			
Playing *Penny Plate* (*Teacher's Lesson Guide*, pages 44 and 45)	**GMP 6.3** Be accurate when you count, measure, and calculate. See also: **GMP 2.1**	Children practice finding sum-equals-ten facts to develop automaticity.	How did you know how many pennies were under the plate? How can you check whether your answer is correct? **Why is it important to count, add, and subtract correctly?**
Lesson 1◆7: Working in Small Groups			
Making a Class Number Scroll from 1 to 1,000 (*Teacher's Lesson Guide*, pages 48 and 49)	**GMP 3.2** Work to make sense of others' mathematical thinking. See also: **GMP 3.1, GMP 6.3, GMP 7.1**	Children work in small groups to create a class number scroll from 1 to 1,000. While working in groups, children discuss rules and patterns represented in the number scroll.	How did your group work together to complete the grid? How did your group decide what number to write in each box? What did you do to understand the mathematical thinking of others in your group? **Why is it important to make sense of others' mathematical thinking?**
Lesson 1◆8: Number Grids			
Finding Patterns on a Number Grid; Completing Number-Grid Puzzles (*Teacher's Lesson Guide*, pages 52 and 53)	**GMP 7.1** Find, extend, analyze, and create patterns. See also: **GMP 6.1, GMP 7.2**	Children identify various number grid patterns, such as 1 more, 10 more, 1 less, and 10 less.	What are some of the patterns that you see on the number grid? How do the patterns on the number grid help you fill in missing numbers on the grid? **Where can you find patterns in mathematics?**
Lesson 1◆9: Equivalent Names for Numbers			
Solving Broken Calculator Problems (*Teacher's Lesson Guide*, page 58)	**GMP 1.4** Solve your problem in more than one way. See also: **GMP 2.1, GMP 5.2, GMP 6.3**	Children find equivalent names for numbers while practicing using calculators.	How are equivalent names for numbers similar? How are they different? How might you use one equivalent name for a number to help you find a different equivalent name for the same number? Why is it important to write names for numbers in more than one way? **When might it be helpful to solve a problem in more than one way? Explain your thinking.**

Grade 2 Unit 1
Numbers and Routines (cont.)

Activity	*Everyday Mathematics* Goal for Mathematical Practice	Opportunity	Guiding Questions
Lesson 1◆10: Counting Patterns			
Counting with a Calculator (*Teacher's Lesson Guide,* pages 62 and 63)	**GMP 3.1** Explain both what you do and why it works. See also: GMP 5.2, GMP 6.3, GMP 7.1, GMP 7.2	Children use calculators to skip count and share patterns that they notice.	What do you do to determine whether a 2-digit number is odd or even? How do you know your strategy works? **Why is it important to explain your strategies and be able to say why they work?**
Lesson 1◆11: Relations (<, >, =) and Home Links			
Reviewing Relations: Less Than (<), Greater Than (>), Equal To (=) (*Teacher's Lesson Guide,* pages 67 and 68)	**GMP 2.2** Explain the meaning of the numbers, words, pictures, symbols, gestures, tables, graphs, and concrete objects you and others use. See also: GMP 2.1	Children compare numbers using words, and read and write number sentences with the relation symbols <, >, and =.	When would you use the symbols < and >? What do the symbols mean? Explain your thinking. When would you use the symbol =? What does it mean? Explain your thinking. When comparing numbers, how do you know which symbols to use? **Why is it important to be able to explain what numbers and symbols mean?**
Lesson 1◆12: Exploring Temperatures, Base-10 Structures, and Dominoes			
Exploration A: Measuring Temperature (*Teacher's Lesson Guide,* page 73)	**GMP 5.2** Use mathematical tools correctly and efficiently. See also: GMP 2.1, GMP 4.1, GMP 6.3	Children read and display temperatures on the class thermometer.	What information does a thermometer provide? Describe how you use a thermometer. **Why is it important to use a thermometer or any other tool correctly?**

Grade 2 Unit 2
Addition and Subtraction Facts

Activity	*Everyday Mathematics* Goal for Mathematical Practice	Opportunity	Guiding Questions
Lesson 2◆1: Addition Number Stories			
Completing Number-Grid Puzzles (*Teacher's Lesson Guide,* page 97)	**GMP 7.1** Find, extend, analyze, and create patterns. See also: **GMP 1.1, GMP 7.2**	Children use patterns to complete number-grid puzzles.	What patterns do you see on the number grid? How do patterns help you fill in missing numbers on the grid? **Where else can you find patterns in mathematics?**
Lesson 2◆2: Review "Easy" Addition Facts			
Reviewing +0 and +1 Shortcuts (*Teacher's Lesson Guide,* pages 100 and 101)	**GMP 8.2** Use properties, rules, and shortcuts to solve problems. See also: **GMP 7.1, GMP 8.1**	Children review, practice, and discuss the +0 and +1 shortcuts.	Give examples of the +0 and +1 shortcuts. How might you explain them to someone? **How can these and other shortcuts help you in mathematics?**
Lesson 2◆3: Doubles Facts			
Reviewing the Facts Table (*Teacher's Lesson Guide,* pages 106 and 107)	**GMP 7.1** Find, extend, analyze, and create patterns. See also: **GMP 5.2**	Children explore patterns on the Facts Table.	What patterns do you notice on the Facts Table? Give an example of one of the patterns on the Facts Table. How might this pattern help you learn your addition facts? **Why is it important to know how to find patterns in mathematics?**
Lesson 2◆4: Turn-Around Facts and the +9 Shortcut			
Introducing the +9 Shortcut (*Teacher's Lesson Guide,* page 113)	**GMP 8.1** Use patterns and structures to create and explain rules and shortcuts. See also: **GMP 7.1**	Children use the Facts Table to find and discuss shortcuts for solving +9 facts.	How might you describe the +9 patterns on the Facts Table? Give an example of a +9 fact. How might you use the +9 shortcut to find the sum? **What other shortcuts in mathematics do you know?**
Lesson 2◆5: Addition Strategies That Use Doubles Facts			
Discussing Doubles-Plus-1 Facts; Discussing Doubles-Plus-2 Facts (*Teacher's Lesson Guide,* pages 117 and 118)	**GMP 8.2** Use properties, rules, and shortcuts to solve problems. See also: **GMP 6.3**	Children discuss and practice strategies for doubles-plus-1 and doubles-plus-2 facts.	Give an example of a doubles-plus-1 fact. How would you use the doubles-plus-1 strategy to find the sum? Give an example of a doubles-plus-2 fact. How would you use the doubles-plus-2 strategy to find the sum? How does knowing doubles facts help you to learn other facts? **How can these and other shortcuts help you in mathematics?**

Grade 2 Unit 2
Addition and Subtraction Facts (cont.)

Activity	*Everyday Mathematics* Goal for Mathematical Practice	Opportunity	Guiding Questions
Lesson 2◆6: Subtraction from Addition			
Using Dominoes to Generate Related Addition and Subtraction Facts (*Teacher's Lesson Guide,* page 123)	**GMP 4.2** Use mathematical models such as graphs, drawings, tables, symbols, numbers, and diagrams to solve problems. *See also:* **GMP 2.1, GMP 2.2, GMP 5.2, GMP 8.2**	Children use dominoes to discover and practice related addition and subtraction facts.	How can you use the number of dots on a domino to make two addition and two subtraction facts? **How does using dominoes and other mathematical models help you in mathematics?**
Lesson 2◆7: Fact Families			
Math Boxes 2◆7 Writing/Reasoning Prompt (*Teacher's Lesson Guide,* page 129)	**GMP 3.1** Explain both what to do and why it works. *See also:* **GMP 2.1, GMP 6.1, GMP 8.1**	Children create a fact family and describe how knowing addition facts helps solve subtraction facts.	How does writing an addition fact for a fact family help you know the related subtraction fact? Explain why this works. **In math, why is it important to explain what you did and why it works?**
Lesson 2◆8: Exploring Weights and Scales, Odd and Even Patterns, and Equal Groups			
Exploration A: Using a Pan Balance and a Spring Scale (*Teacher's Lesson Guide,* page 134)	**GMP 1.2** Make a plan for solving your problem. *See also:* **GMP 1.1, GMP 1.3, GMP 1.5, GMP 2.1, GMP 4.1, GMP 5.2, GMP 6.2**	Children use a pan balance and spring scale to compare the weight of objects and select objects that weigh about 1 pound.	What was your plan for choosing objects that weighed about 1 pound? **Why is it important to make a plan before attempting to solve a problem?**
Lesson 2◆9: Name Collections			
Math Message Follow-Up; Practicing with Name-Collection Boxes (*Teacher's Lesson Guide,* pages 139 and 141)	**GMP 1.4** Solve your problem in more than one way. *See also:* **GMP 2.1, GMP 6.3**	Children use name-collection boxes to create equivalent names for numbers.	How are other names for the same number similar? How are they different? How might you use one name for a number to help you find a different name for the same number? Why is it important to write names for numbers in more than one way? **When might it be helpful to solve your problem in more than one way? Explain your thinking.**

Grade 2 Unit 2
Addition and Subtraction Facts (cont.)

Activity	*Everyday Mathematics* Goal for Mathematical Practice	Opportunity	Guiding Questions
Lesson 2•10: Frames-and-Arrows Routines			
Demonstrating Frames-and-Arrows Routines (*Teacher's Lesson Guide*, pages 144–146)	GMP 7.1 Find, extend, analyze, and create patterns. See also: GMP 1.1, GMP 1.6, GMP, 2.1, GMP 6.3, GMP 8.1, GMP 8.2	Children review Frames-and-Arrows routines and use addition and subtraction rules to generate patterns.	What is a pattern? How do you find the pattern in a Frames-and-Arrows problem? How do you know what number to write in each empty frame? **Why are patterns important in mathematics?**
Lesson 2•11: "What's My Rule?" Routines			
Using Function Machines to Illustrate "What's My Rule?" Tables (*Teacher's Lesson Guide*, pages 151 and 152)	GMP 8.2 Use properties, rules, and shortcuts to solve problems. See also: GMP 1.1, GMP 1.2, GMP 2.1, GMP 6.3, GMP 7.1, GMP 8.1	Children use inputs, outputs, and and/or rules to complete "What's My Rule?" problems.	What is the relationship between the *in* and *out* numbers and the rule in a *"What's My Rule?"* problem? What information can you use to find a missing rule in a *"What's My Rule?"* problem? **How are rules used in mathematics?**
Lesson 2•12: Counting Strategies for Subtraction			
Reviewing the Counting-Back Strategy for Subtraction and Reviewing the Counting-Up Strategy for Subtraction (*Teacher's Lesson Guide*, pages 156 and 157)	GMP 6.3 Be accurate when you count, measure, and calculate. See also: GMP 7.1	Children review the counting-back and counting-up strategies and use them to practice basic subtraction problems.	How did you know whether your differences were correct? **Why is it important to be accurate when you calculate?**
Lesson 2•13: Shortcuts for "Harder" Subtraction Facts			
Introducing the –9 Shortcut and Introducing the –8 Shortcut (*Teacher's Lesson Guide*, pages 162 and 163)	GMP 8.2 Use properties, rules, and shortcuts to solve problems. See also: GMP 6.3	Children discuss and practice –9 and –8 shortcuts.	Give an example of a –9 fact. How would you use the –9 shortcut to solve a subtraction problem? Give an example of a –8 fact. How would you use the –8 shortcut to solve a subtraction problem? **How can these and other shortcuts help you in mathematics?**

Grade 3 Unit 1
Routines, Review, and Assessment

Activity	*Everyday Mathematics* Goal for Mathematical Practice	Opportunity	Guiding Questions
Lesson 1◆1: Numbers and Number Sequences			
Completing Number Sequences (*Teacher's Lesson Guide,* page 20)	**GMP 6.3** Be accurate when you count, measure, and calculate. See also: **GMP 1.1, GMP 7.1**	Children complete partial sequences with three- and four-digit numbers in both verbal and written form.	When completing a number sequence, how do you know which number comes next? How do you know your counting is accurate? **Why is it important to count accurately?**
Lesson 1◆2: Number Grids			
Reviewing Number-Grid Patterns (*Teacher's Lesson Guide,* pages 24 and 25)	**GMP 7.1** Find, extend, analyze, and create patterns. See also: **GMP 3.2, GMP 5.2**	Children identify patterns on the number grid.	What are some of the patterns that you see on the number grid? How do the patterns on the number grid help you fill in missing numbers on the grid? **Where can you find patterns in mathematics?**
Lesson 1◆3: Introducing the *Student Reference Book*			
Playing *Less Than You!* (*Teacher's Lesson Guide,* page 31)	**GMP 6.3** Be accurate when you count, measure, and calculate.	Children practice mental addition skills and develop a winning game strategy by playing *Less Than You!*	How do you know that your sums are correct? How do you know that your partner's sums are correct? **Why is it important to be accurate when you add, subtract, multiply, or divide?**
Lesson 1◆4: Tools for Mathematics			
Using Mathematical Tools (*Teacher's Lesson Guide,* page 34)	**GMP 5.2** Use mathematical tools correctly and efficiently. See also: **GMP 6.3, GMP 2.1**	Children use a variety of mathematical tools to solve problems.	Select a tool and describe how it helped you solve a problem. **Why is it important to use mathematical tools correctly?**
Lesson 1◆5: Analyzing and Displaying Data			
Making Tally Charts and Bar Graphs (*Teacher's Lesson Guide,* pages 39 and 40)	**GMP 2.2** Explain the meaning of the numbers, words, pictures, symbols, gestures, tables, graphs, and concrete objects you and others use. See also: **GMP 2.1, GMP 4.1**	Children display data in a tally chart and a bar graph. Then they find the landmarks (maximum, minimum, range, mode, and median) of the data.	What does each tally mark in the tally chart mean? What does each square in the bar graph mean? What are the advantages of displaying data in a tally chart or a bar graph? **Why is it important to be able to explain data shown in tally charts and graphs?**

Grade 3 Unit 1
Routines, Review, and Assessment (cont.)

Activity	*Everyday Mathematics* Goal for Mathematical Practice	Opportunity	Guiding Questions
Lesson 1◆6: Equivalent Names			
Completing Name-Collection Box Problems (*Teacher's Lesson Guide,* page 44)	**GMP 1.4** Solve your problem in more than one way. *See also:* **GMP 2.1, GMP 6.3**	Children generate equivalent mathematical expressions in Name-Collection Box problems.	How are equivalent names for numbers similar? How are they different? How might you use one equivalent name for a number to help you find a different equivalent name for the same number? Why is it important to write names for numbers in more than one way? **When might it be helpful to solve a problem in more than one way? Explain your thinking.**
Lesson 1◆7: The Language of Chance Events			
Introducing Words and Phrases Associated with Chance Events; Making Lists of Certain and Uncertain Events (*Teacher's Lesson Guide,* pages 48 and 49)	**GMP 4.1** Apply mathematical ideas to real-world situations. *See also:* **GMP 2.1**	Children discuss words related to chance events, including: sure will happen, sure will not happen, likely, unlikely, 50-50 chance, and so on. They make lists of events that they are sure will happen, sure will not happen, and think might happen.	How do you decide whether an event is sure to happen, sure not to happen, or might happen? Why is it important to know whether events are sure to happen, sure not to happen, or might happen? **When might you use these terms in real life?**
Lesson 1◆8: Finding Differences			
Finding Differences (*Teacher's Lesson Guide,* pages 52 and 53)	**GMP 6.3** Be accurate when you count, measure, and calculate. *See also:* **GMP 3.1, GMP 3.2, GMP 5.2, GMP 6.1**	Children use a number grid to find the difference between two numbers and explain the strategies they used.	How can you be sure that you are counting accurately on the number grid when finding the difference between two numbers? How do you know whether the difference you find is correct? **When solving any problem, how do you know if your answer is correct?**
Lesson 1◆9: Calculator Routines			
Practicing Calculator Skills (*Teacher's Lesson Guide,* pages 57 and 58)	**GMP 5.2** Use mathematical tools correctly and efficiently. *See also:* **GMP 1.1, GMP 6.3, GMP 7.1, GMP 7.2**	Children use calculators to solve number stories and change-to problems, problems involving multiple addends and 4-digit by 4-digit subtraction, and to skip count.	How do you decide whether or not to use a calculator to solve a problem? **Why is it important to use a calculator, or any other tool, correctly?**

Grade 3 Unit 1
Routines, Review, and Assessment (cont.)

Activity	*Everyday Mathematics* Goal for Mathematical Practice	Opportunity	Guiding Questions
Lesson 1♦10: Money			
Practicing Skills with Money (*Teacher's Lesson Guide*, page 64)	**GMP 2.2** Explain the meaning of the numbers, words, pictures, symbols, gestures, tables, graphs, and concrete objects you and others use. See also: GMP 1.1, GMP 2.1, GMP 4.1, GMP 6.1, GMP 6.3	Children practice dollar-and-cents notation, compare money amounts using the symbols < and >, and make change.	Why is the decimal point important when solving problems with money? When comparing money amounts, how do you know which symbols to use? **Why is it important for you to be able to explain what numbers and symbols mean?**
Lesson 1♦11: Solving Problems with Dollars and Cents			
Math Message Follow-Up (*Teacher's Lesson Guide*, pages 68–70)	**GMP 6.2** Use the level of precision you need for your problem. See also: GMP 2.1, GMP 3.1, GMP 3.2, GMP 4.1, GMP 6.3	Children solve money problems and discuss their solution strategies. They also discuss situations where estimates are more appropriate than exact answers.	What is the difference between an estimated answer and an exact answer? **How do you decide whether the answer to a problem should be exact or an estimate?**
Lesson 1♦12: Patterns			
Reviewing Frames-and-Arrows Routines (*Teacher's Lesson Guide*, pages 74–76)	**GMP 7.1** Find, extend, analyze, and create patterns. See also: GMP 1.1, GMP 6.3, GMP 8.1, GMP 8.2	Children identify patterns to complete Frames-and-Arrows diagrams.	What is a pattern? How do you find the pattern in a Frames-and-Arrows problem? How do you know what number to write in each empty frame? **Why are patterns important in mathematics?**
Lesson 1♦13: The Length-of-Day Project			
Math Message Follow-Up (*Teacher's Lesson Guide*, page 79)	**GMP 6.1** Communicate your mathematical thinking clearly and precisely. See also: GMP 1.1, GMP 2.1, GMP 4.1, GMP 6.3	Children solve problems involving elapsed time and share solution strategies.	Explain your solution to the problem. Show what you did. **Why is it important to communicate your mathematical thinking clearly?**

Grade 3 Unit 2
Adding and Subtracting Whole Numbers

Activity	*Everyday Mathematics* Goal for Mathematical Practice	Opportunity	Guiding Questions
Lesson 2◆1: Fact Families			
Reviewing Fact Family Concepts (*Teacher's Lesson Guide,* pages 101 and 102)	**GMP 3.1** Explain both what to do and why it works. *See also:* **GMP 8.2**	Children review the turn-around rule (the Commutative Property of Addition) and the relationship between addition and subtraction for a fact family.	Give an example of a fact family. How did you decide which numbers to use for the addition and subtraction facts in your family? Why do turn-around facts work for addition but not for subtraction? How would you explain the relationship between addition and subtraction? **Why is it important to be able to explain what you did to solve a math problem and why it works?**
Lesson 2◆2: Extensions of Addition and Subtraction Facts			
Practicing Fact Extensions (*Teacher's Lesson Guide,* page 108)	**GMP 7.2** Use patterns and structures to solve problems. *See also:* **GMP 5.2, GMP 6.3, GMP 7.1**	Children use basic addition and subtraction facts to solve fact extensions.	How does knowing a basic fact help you to solve problems with larger numbers? What is the relationship between ones, tens, and hundreds? **When is the relationship between ones, tens, and hundreds important in mathematics?** **How does knowing about the relationship between ones, tens, and hundreds help you solve problems with larger numbers?**
Lesson 2◆3: "What's My Rule?"			
Completing "What's My Rule?" Tables (*Teacher's Lesson Guide,* page 114)	**GMP 8.1** Use patterns and structures to create and explain rules and shortcuts. *See also:* **GMP 1.1, GMP 2.1, GMP 6.3, GMP 7.1, GMP 8.2**	Children review and solve variations of "What's My Rule?" problems.	What is the relationship between *in* and *out* numbers and the rule in a "What's My Rule?" problem? What information might you use to find a rule that is unknown in a "What's My Rule?" problem? **Why are rules important in mathematics?**

Grade 3 Unit 2
Adding and Subtracting Whole Numbers (cont.)

Activity	*Everyday Mathematics* Goal for Mathematical Practice	Opportunity	Guiding Questions
Lesson 2◆4: Parts-and-Total Number Stories			
Using the Guide to Solving Number Stories (*Teacher's Lesson Guide,* pages 118 and 119)	**GMP 1.2** Make a plan for solving your problem. *See also:* **GMP 1.1, GMP 1.5, GMP 2.1, GMP 3.1, GMP 4.1, GMP 4.2, GMP 6.3**	Children use the Guide to Solving Number Stories and parts-and-total situation diagrams to solve number stories.	Describe a time when you made a plan to do something at school or at home. How was making that plan similar to making a plan to solve a math problem? **Why is it important to make a plan when solving math problems?**
Lesson 2◆5: Change Number Stories			
Solving a Change-to-More Number Story; Solving a Change-to-Less Number Story (*Teacher's Lesson Guide,* pages 124–126)	**GMP 4.2** Use mathematical models such as graphs, drawings, tables, symbols, numbers, and diagrams to solve problems. *See also:* **GMP 1.1, GMP 1.2, GMP 1.5, GMP 2.1, GMP 3.1, GMP 6.1, GMP 6.3**	Children use change diagrams to solve change-to-more and change-to-less number stories.	How do you know what numbers to write in the diagram and where to write them? How do you know what numbers and symbols to write in your number model? **How can diagrams and number models help you solve math problems?**
Lesson 2◆6: Comparison Number Stories			
Introducing the National High/Low Temperatures Project (*Teacher's Lesson Guide,* page 133)	**GMP 4.1** Apply mathematical ideas to real-world situations. *See also:* **GMP 2.1, GMP 3.2, GMP 5.2, GMP 6.3**	Children begin the National High/Low Temperatures Project by comparing and recording high and low temperatures.	Why might it be helpful to know high and low temperatures? **Why is collecting, comparing, and recording data important?**
Lesson 2◆7: The Partial-Sums Algorithm			
Making Ballpark Estimates (*Teacher's Lesson Guide,* page 136)	**GMP 1.5** Check whether your solution makes sense. *See also:* **GMP 6.2, GMP 7.2**	Children use ballpark estimates to judge the accuracy of their solutions to addition problems.	Why are ballpark estimates a good way to check whether your answer makes sense? How do you know whether your answer is correct? **Why is it important to check whether your answer makes sense?**

Grade 3 Unit 2
Adding and Subtracting Whole Numbers (cont.)

Activity	*Everyday Mathematics* Goal for Mathematical Practice	Opportunity	Guiding Questions
Lesson 2♦8: Subtraction Algorithms			
Math Boxes, Writing/Reasoning Response (*Teacher's Lesson Guide*, page 145)	**GMP 4.1** Apply mathematical ideas to real-world situations. See also: **GMP 2.1, GMP 3.1, GMP 6.3**	Children find the time on an analog clock and say what time it will be in 30 minutes from that time.	What are some things that you do at 2:25 A.M. and 2:25 P.M.? **When might you need to know what time it will be in 30 minutes?** Why is telling time important?
Lesson 2♦9: Addition with Three or More Addends			
Solving Number Stories Having Three or More Addends (*Teacher's Lesson Guide*, page 150)	**GMP 1.1** Work to make sense of your problem. See also: **GMP 1.2, GMP 1.5, GMP 2.1, GMP 3.1, GMP 4.2, GMP 6.3**	Children use parts-and-total diagrams to solve number stories with three and four addends, and share their solutions and strategies.	What questions can you ask yourself to help make sense of a number story? **Why is it important to make sense of problems in mathematics?**

Grade 4 Unit 1
Naming and Constructing Geometric Figures

Activity	*Everyday Mathematics* Goal for Mathematical Practice	Opportunity	Guiding Questions
Lesson 1♦1: Introduction to the *Student Reference Book*			
Investigating the *Student Reference Book* (*Teacher's Lesson Guide,* pages 19 and 20)	**GMP 5.2** Use mathematical tools correctly and efficiently. See also: **GMP 6.1**	Students explore the *Student Reference Book,* a resource tool they will use throughout the year.	How is this book organized? How can this book help you with your homework? **How can this tool help you work more efficiently?**
Lesson 1♦2: Points, Line Segments, Lines, and Rays			
Reviewing Points, Line Segments, Lines, and Rays (*Teacher's Lesson Guide,* pages 25 and 26)	**GMP 2.2** Explain the meaning of the numbers, words, pictures, symbols, gestures, tables, graphs, and concrete objects you and others use. See also: **GMP 1.6, GMP 2.1, GMP 5.2, GMP 6.1**	Students review the definitions, characteristics, and symbols for points, line segments, lines, and rays.	How are a line segment, a line, and a ray different? How are they similar? **How does explaining a term help you understand it better?**
Lesson 1♦3: Angles, Triangles, and Quadrangles			
Constructing Triangles and Quadrangles (*Teacher's Lesson Guide,* pages 31 and 32)	**GMP 2.1** Represent problems and situations mathematically with numbers, words, pictures, symbols, gestures, tables, graphs, and concrete objects. See also: **GMP 2.2, GMP 6.1, GMP 7.1, GMP 7.2**	Students use straws to display and study three- and four-sided shapes.	What is the minimum number of angles needed to make a shape? How can you use straws to prove your answer? Why can a shape have more than one name? **How do straw representations help you see the characteristics of different shapes?**

Grade 4 Unit 1
Naming and Constructing Geometric Figures (cont.)

Activity	*Everyday Mathematics* Goal for Mathematical Practice	Opportunity	Guiding Questions
Lesson 1◆4: Parallelograms			
Exploring Parallelograms (*Teacher's Lesson Guide*, page 38)	**GMP 8.2** Use properties, rules, and shortcuts to solve problems. *See also:* GMP 2.1, GMP 5.1, GMP 5.2, GMP 6.1, GMP 8.1	Students discuss relationships, similarities, and differences of quadrangles.	What is a property? What are some properties of quadrangles? **How did looking at similarities and differences among quadrangles help you categorize the shapes?** **How can using properties help you solve problems?**
Lesson 1◆5: Polygons			
Defining the Properties of Polygons (*Teacher's Lesson Guide*, page 44)	**GMP 8.1** Use patterns and structures to create and explain rules and shortcuts. *See also:* GMP 1.1, GMP 1.6, GMP 3.2, GMP 6.1, GMP 7.1, GMP 8.2	Partners compare and contrast figures to develop a definition for a polygon.	What are the properties of a polygon? How did the examples help you determine the properties of a polygon? **Why is it important to determine properties of shapes?**
Lesson 1◆6: Drawing Circles with a Compass			
Drawing Circles with a Compass (*Teacher's Lesson Guide*, pages 48 and 49)	**GMP 5.2** Use mathematical tools correctly and efficiently. *See also:* GMP 1.4	Students experiment with different ways to use a compass to construct circles.	How did trying different methods help you find a comfortable way to use your compass? **Why is it important to practice using a tool correctly?** **How do tools help you work more efficiently?**

Grade 4 Unit 1
Naming and Constructing Geometric Figures (cont.)

Activity	*Everyday Mathematics* Goal for Mathematical Practice	Opportunity	Guiding Questions
Lesson 1•7: Circle Constructions			
Math Message Follow-Up (*Teacher's Lesson Guide,* page 53)	**GMP 1.6** Connect mathematical ideas and representations to one another. See also: **GMP 2.1, GMP 2.2, GMP 3.2, GMP 5.1, GMP 5.2, GMP 6.2**	Partners make 20 points equidistant from a given point to discover the properties of a circle and its radius.	What did you notice when measuring the radius a second time? **Why is it important to understand the connection between the points, the circle, and its radius?** Why is it important to connect math ideas to each another?
Lesson 1•8: Hexagon and Triangle Constructions			
Making Constructions with a Compass and Straightedge (*Teacher's Lesson Guide,* pages 58 and 59)	**GMP 6.2** Use the level of precision you need for your problem. See also: **GMP 2.1, GMP 5.2, GMP 6.1, GMP 6.2, GMP 7.1**	Students construct an inscribed, regular hexagon, and divide it into 6 equilateral triangles.	Explain how you know that all the vertices of the hexagon are the same distance from the center of the circle? Why do you need to be precise when creating your hexagon? **Give an example of a real-life situation where precision is needed, and explain why it is necessary.**

43

Grade 4 Unit 2
Using Numbers and Organizing Data

Activity	*Everyday Mathematics* Goal for Mathematical Practice	Opportunity	Guiding Questions
Lesson 2◆1: A Visit to Washington, D.C.			
Examining Numerical Information about Washington, D.C. (*Teacher's Lesson Guide,* pages 85 and 86)	**GMP 2.2** Explain the meanings of the numbers, words, pictures, symbols, gestures, tables, graphs, and concrete objects you and others use. *See also:* GMP 4.1, GMP 4.2, GMP 5.1, GMP 6.2	Students use numerical data about Washington, D.C., from the *Student Reference Book* to explore uses of numbers.	What kinds of units did you find in the essay? How did the units help you figure out whether a number was a count, code, measure, etc.? **Why is it important to understand what numbers mean? What would happen if we didn't have numbers?**
Lesson 2◆2: Many Names for Numbers			
Reviewing the Idea that Numbers Have Many Names (*Teacher's Lesson Guide,* pages 90 and 91)	**GMP 1.4** Solve your problem in more than one way. *See also:* GMP 1.6, GMP 2.1, GMP 3.2, GMP 6.3, GMP 7.1, GMP 7.2	Students generate a variety of mathematical expressions equivalent to a given number.	How are these equivalent names for _____ similar? How are they different? How can you use one equivalent name to help you find another equivalent name? **How is it helpful to solve a problem in more than one way?**
Lesson 2◆3: Place Value in Whole Numbers			
Reviewing Place Value for Whole Numbers (*Teacher's Lesson Guide,* pages 95 and 96)	**GMP 7.1** Find, extend, analyze, and create patterns. *See also:* GMP 2.2, GMP 6.1, GMP 8.1	Students use patterns and structures in our place-value number system to read and write numbers.	What patterns do you see in how we write numbers? What patterns do you see in how we say numbers? **Why is our number system called Base-10? How can just 10 digits form all the whole numbers there are?**
Lesson 2◆4: Place Value with a Calculator			
Practicing Place-Value Skills with a Calculator (*Teacher's Lesson Guide,* pages 102 and 103)	**GMP 7.2** Use patterns and structures to solve problems. *See also:* GMP 1.2, GMP 5.2	Students use patterns and structures in our place-value number system to solve problems.	How can place-value patterns help you figure out how to change the numbers? What patterns are most helpful in solving these problems? **How can a pattern help you solve a problem?**

Grade 4 Unit 2
Using Numbers and Organizing Data (cont.)

Activity	*Everyday Mathematics* Goal for Mathematical Practice	Opportunity	Guiding Questions
Lesson 2♦5: Organizing and Displaying Data			
Collecting, Organizing, and Interpreting a Set of Data (*Teacher's Lesson Guide,* pages 107–109)	**GMP 2.1** Represent problems and situations mathematically with numbers, words, pictures, symbols, gestures, tables, graphs, and concrete objects. See also: **GMP 1.6, GMP 2.2, GMP 3.1, GMP 4.1, GMP 4.2, GMP 5.1, GMP 6.2**	Students create a tally chart and find statistical landmarks to organize and describe data from boxes of raisins.	How does the tally chart help you learn about the number of raisins in the boxes? What other ways could you organize the data? **Why is it important to organize data?**
Lesson 2♦6: The Median			
Investigating the Sizes of Students' Families (*Teacher's Lesson Guide,* pages 113–116)	**GMP 2.2** Explain the meanings of the numbers, words, pictures, symbols, gestures, tables, graphs, and concrete objects you and others use. See also: **GMP 4.1, GMP 4.2, GMP 5.1**	Students use a line plot and statistical landmarks to organize, display, and understand data about family size.	How does it help to see the data in a line plot? Where is the median on the line plot? What does the median tell us about the typical family size? **Why is it important to understand what numbers and graphs mean?**
Lesson 2♦7: Addition of Multidigit Numbers			
Math Boxes 2♦7 Writing/Reasoning Prompt (*Teacher's Lesson Guide,* page 123)	**GMP 3.2** Work to make sense of others' mathematical thinking. See also: **GMP 3.1, GMP 6.1**	Students respond in writing to a claim about what geometric figures fit the definition of a parallelogram.	What might have made Shaneel think the way he does? What should you do when you see a solution you think is incorrect? **Why is it important to question an answer you think is incorrect? Why is it important to make sense of other people's mathematical thinking?**

Grade 4 Unit 2
Using Numbers and Organizing Data (cont.)

Activity	*Everyday Mathematics* Goal for Mathematical Practice	Opportunity	Guiding Questions
Lesson 2◆8: Displaying Data with Graphs			
Collecting and Organizing Head-Size Data (*Teacher's Lesson Guide*, pages 127–129)	**GMP 4.2** Use mathematical models such as graphs, drawings, tables, symbols, numbers, and diagrams to solve problems. See also: **GMP 1.6, GMP 2.1, GMP 2.2, GMP 4.1**	Students approach a problem about hat sizes by mathematically modeling their own head sizes. They measure their head sizes, find the median, and make a bar graph and line plot of the data.	How do these models help you make a recommendation about hat sizes? What do the landmarks tell you about the hat sizes? What are some other models that you could use to help you solve the problem? **Why is it useful to graph your data? How can mathematical models help you solve problems?**
Lesson 2◆9: Subtraction of Multidigit Numbers			
Discussing and Practicing the Partial-Differences Method for Subtraction (*Teacher's Lesson Guide*, pages 135 and 136)	**GMP 3.1** Explain both what to do and why it works. See also: **GMP 3.2, GMP 5.2, GMP 5.3, GMP 6.3, GMP 8.2**	Students explain how to solve problems using the partial-differences method for subtraction.	How would you explain this method to someone who doesn't know it? What tools could you use to explain this method? How does this method compare to other subtraction methods you know? **Why is it important for you to explain how you solve problems?**

Grade 5 Unit 1 Number Theory

Activity	*Everyday Mathematics* Goal for Mathematical Practice	Opportunity	Guiding Questions
Lesson 1♦1: Introduction to the *Student Reference Book*			
Solving Problems Using the *Student Reference Book* (*Teacher's Lesson Guide*, page 18)	**GMP 5.2** Use mathematical tools correctly and efficiently. See also: **GMP 3.1**	Students use the Table of Contents, Glossary, and Index of the *Student Reference Book* to compete in a mathematical scavenger hunt.	How will the organization of this book help you to find information? How can this book help you with your homework? **How will this tool help you work more efficiently?**
Lesson 1♦2: Rectangular Arrays			
Recognizing Patterns in Extended Facts (*Teacher's Lesson Guide*, page 25)	**GMP 8.1** Use patterns and structures to create and explain rules and shortcuts. See also: **GMP 3.1, GMP 7.1, GMP 7.2, GMP 8.2**	Students describe patterns for finding products and quotients with extended facts.	How does knowing basic facts make solving extended facts easier? What did you notice about the patterns in the product when you multiplied a 2-digit number by a 2-digit number? What is the rule for solving extended facts? **How can patterns help you solve problems and explain rules?**
Lesson 1♦3: Factors			
Finding Factor Pairs (*Teacher's Lesson Guide*, pages 29 and 30)	**GMP 2.1** Represent problems and situations mathematically with numbers, words, pictures, symbols, gestures, tables, graphs, and concrete objects. See also: **GMP 1.2, GMP 1.6, GMP 4.2**	Students create arrays and write number models to find all the possible factors of given numbers.	Is there another representation that would help you find all of the factors of a given number? **How does representing a mathematical situation with words or through visuals increase your understanding of a problem? How do mathematical symbols such as +, *, and = help you represent your problem?**
Lesson 1♦4: The *Factor Captor* Game			
Playing *Factor Captor* (*Teacher's Lesson Guide*, pages 34 and 35)	**GMP 3.1** Explain both what to do and why it works. See also: **GMP 1.2, GMP 3.2, GMP 6.1**	Students practice finding factors for whole numbers when they play *Factor Captor*.	Explain the reasoning for your lead number and why you think it is the next best move. **Why is it important to explain what you are doing and why it works?**

Grade 5 Unit 1
Number Theory (cont.)

Activity	*Everyday Mathematics* Goal for Mathematical Practice	Opportunity	Guiding Questions
Lesson 1◆5: Divisibility			
Introducing Divisibility Rules (*Teacher's Lesson Guide,* page 39)	**GMP 8.2** Use properties, rules, and shortcuts to solve problems. See also: GMP 1.6, GMP 3.2, GMP 5.2, GMP 8.1	Students apply divisibility rules for 1, 2, 3, 5, 6, 9, and 10.	How can we check to see if the rules work? How can divisibility rules be useful in real life? **How can mathematical rules and shortcuts help you to become a stronger mathematical thinker?**
Lesson 1◆6: Prime and Composite Numbers			
Developing a Strategy for *Factor Captor* (*Teacher's Lesson Guide,* pages 44 and 45)	**GMP 3.1** Explain both what to do and why it works. See also: GMP 1.1, GMP 1.2, GMP 3.2, GMP 6.1, GMP 8.1, GMP 8.2	Students study prime and composite numbers to develop a strategy for playing *Factor Captor*.	How would you describe your strategy? Explain why you believe your strategy is effective. How does understanding prime and composite numbers help you make the best selection for your numbers? **What is an example of a strategy you could use every time you play the game? Explain it.**
Lesson 1◆7: Square Numbers			
Investigating the Properties of Square Numbers (*Teacher's Lesson Guide,* pages 49 and 50)	**GMP 6.1** Communicate your mathematical thinking clearly and precisely. See also: GMP 2.1, GMP 3.1, GMP 5.2, GMP 7.1, GMP 8.1, GMP 8.3	Students are introduced to exponential notation and the calculator exponent key as they investigate the properties of square numbers.	Do you think the pattern of odd and even square numbers continues after 100? How could you test it? How is squaring a number different from doubling a number? **What are some of the benefits of using precise and accurate language to communicate your thinking?**
Lesson 1◆8: Unsquaring Numbers			
"Unsquaring" Numbers (*Teacher's Lesson Guide,* pages 53 and 54)	**GMP 1.5** Check whether your solution makes sense. See also: GMP 1.2, GMP 1.4, GMP 3.1, GMP 3.2, GMP 5.1, GMP 5.3, GMP 6.1	Students develop strategies to "unsquare" (find the square root of) a square number without a calculator.	How could the guess-and-check strategy be used for finding the square root of a number? **How do you know if your answer is reasonable? How can checking whether your solution makes sense help you problem solve?**

Grade 5 Unit 1
Number Theory (cont.)

Activity	*Everyday Mathematics* Goal for Mathematical Practice	Opportunity	Guiding Questions
Lesson 1◆9: Factor Strings and Prime Factorizations			
Finding the Prime Factorization Using Factor Trees (*Teacher's Lesson Guide,* page 60)	**GMP 1.4** Solve your problem in more than one way. *See also:* **GMP 2.1, GMP 2.2**	Students construct factor trees to organize their progress when finding all the prime factors of a whole number.	Why might your solution pathway look different from others? Can you describe another way to find the prime factorization? **Why is it important to be flexible in the way you solve a problem?** **How can solving a problem in more than one way help you find the best strategy for you?**

Grade 5 Unit 2
Estimation and Computation

Activity	*Everyday Mathematics* Goal for Mathematical Practice	Opportunity	Guiding Questions
Lesson 2♦1: Estimation Challenge			
Introducing the Estimation Challenge Problem (*Teacher's Lesson Guide*, pages 82 and 83)	**GMP 1.1** Work to make sense of your problem. See also: GMP 1.1, GMP 1.2, GMP 1.3, GMP 2.1, GMP 3.1, GMP 3.2, GMP 4.1, GMP 5.2, GMP 5.3, GMP 6.1, GMP 6.2, GMP 6.3, GMP 8.3	Students solve a multi-step problem involving time and distance.	Will each group have the same estimate? What are you being asked to solve? What questions need to be answered before the solution can be found? What necessary information do you need to gather to make sense of this problem?
Lesson 2♦2: Addition of Whole Numbers and Decimals			
Math Message Follow-Up (*Teacher's Lesson Guide*, pages 86 and 87)	**GMP 7.1** Find, extend, analyze, and create patterns. See also: GMP 1.4, GMP 1.6, GMP 2.1, GMP 2.2, GMP 3.1, GMP 3.2, GMP 5.1, GMP 6.1, GMP 7.2, GMP 8.2	Students study place-value patterns in whole and decimal numbers.	Why is the digit *zero* important in explaining the patterns in our base-10 number system? Why can't you add clock time numbers the same way you add whole numbers and decimals in our base-10 number system? How are the patterns in the two systems different? How do place-value patterns help you compare and order numbers? What patterns can you describe in the base-10 number system?
Lesson 2♦3: Subtraction of Whole Numbers and Decimals			
Subtracting Whole Numbers and Decimals (*Teacher's Lesson Guide*, page 95)	**GMP 3.1** Explain both what to do and why it works. See also: GMP 1.4, GMP 2.1, GMP 3.2, GMP 5.1, GMP 5.2, GMP 5.3, GMP 6.1, GMP 6.3, GMP 8.2	Students explain and demonstrate two different algorithms for subtracting decimal numbers.	Explain the algorithm you used for one of your problems and why it works. Why is it important to be able to understand and explain how an algorithm works?
Lesson 2♦4: Addition and Subtraction Number Stories			
Playing *Name That Number* (*Teacher's Lesson Guide*, page 101)	**GMP 1.4** Solve your problem in more than one way. See also: GMP 1.2, GMP 1.4, GMP 2.2, GMP 3.1, GMP 3.2, GMP 6.1, GMP 6.3	Students generate many names for a given target number when they play *Name That Number*.	What are some strategies you could consider when you are arranging your cards to name the target number? Why is it important to think about several combinations of the cards in order to generate the target number? Why is it useful to find multiple solutions to a problem?

Grade 5 Unit 2
Estimation and Computation (cont.)

Activity	*Everyday Mathematics* Goal for Mathematical Practice	Opportunity	Guiding Questions
Lesson 2◆5: Estimate Your Reaction Time			
Sharing Results (*Teacher's Lesson Guide*, page 107)	**GMP 2.2** Explain the meanings of the numbers, words, pictures, symbols, gestures, tables, graphs, and concrete objects you and others use. See also: **GMP 1.6, GMP 6.1, GMP 8.3**	Students share their analyses of data collected from an experiment comparing their own left- and right-handed reaction times.	What conclusions can you draw from this experiment? Which data landmark(s) best represents the results of the experiment? Why is it important to analyze and understand your data before you reach a conclusion? What words, objects, or displays can you use to make your explanation clearer?
Lesson 2◆6: Chance Events			
Estimating the Chance That a Thumbtack Will Land Point Down (*Teacher's Lesson Guide*, pages 112 and 113)	**GMP 4.2** Use mathematical models such as graphs, drawings, tables, symbols, numbers, and diagrams to solve problems. See also: **GMP 1.2, GMP 1.6, GMP 2.1, GMP 2.2, GMP 3.2, GMP 4.1, GMP 5.2, GMP 5.3, GMP 6.1**	Students gather and post data on a Probability Meter to figure the chances a dropped thumbtack will land point up or point down.	How does the Probability Meter help you to describe the results of the experiment? How does the position of the stick-on notes on the Probability Meter help you to describe the probability of the tack landing point up or point down? Why is it helpful to use a table to organize and display the results of an experiment? Why are labels and titles important to use on mathematical models? Why is it helpful to use mathematical models to organize and display your information?
Lesson 2◆7: Estimating Products			
Estimating Products (*Teacher's Lesson Guide*, pages 116–118)	**GMP 6.2** Use the level of precision you need for your problem. See also: **GMP 1.4, GMP 1.5, GMP 3.2, GMP 5.3, GMP 6.1**	Students practice strategies for making magnitude estimates for products.	Explain how you determined the most appropriate magnitude estimate for each problem. How does rounding affect your level of precision? How can you use a magnitude estimate to check computation? How can you determine the precision needed to solve a problem?

Grade 5 Unit 2
Estimation and Computation (cont.)

Activity	*Everyday Mathematics* Goal for Mathematical Practice	Opportunity	Guiding Questions
Lesson 2◆8: Multiplication of Whole Numbers and Decimals			
Practicing Multiplication of Whole Numbers and Decimals (*Teacher's Lesson Guide*, pages 123 and 124)	**GMP 5.3** Estimate and use what you know to check the answers you find using tools. See also: GMP 2.1, GMP 3.1, GMP 3.2, GMP 5.1, GMP 5.3, GMP 6.1, GMP 6.2, GMP 6.3, GMP 8.2	Students use both magnitude and ballpark estimates to monitor multiplication answers involving decimals.	What is the difference between a magnitude estimate and a ballpark estimate? How can a magnitude estimate help you place the decimal point in the product? **How does estimation help you check your answer to a multiplication problem?**
Lesson 2◆9: The Lattice Method of Multiplication			
Reviewing the Lattice Method of Multiplication (*Teacher's Lesson Guide*, pages 127–129)	**GMP 5.2** Use mathematical tools correctly and efficiently. See also: GMP 1.5, GMP 5.2, GMP 5.3, GMP 6.1, GMP 6.2, GMP 6.3, GMP 8.2	Students review the lattice method for multiplication and extend the algorithm to include decimals.	How do you know where to place the decimal point when using the lattice method? What are some advantages and disadvantages of the lattice method for multiplication? **Why is it important to use a mathematical tool correctly?** Are there any other mathematical tools that you could use to multiply two numbers?
Lesson 2◆10: Comparing Millions, Billions, and Trillions			
Solving a Tapping Problem with Sampling Strategies (*Teacher's Lesson Guide*, pages 133 and 134)	**GMP 1.2** Make a plan for solving your problem. See also: GMP 1.1, GMP 1.4, GMP 3.1, GMP 4.1, GMP 5.3, GMP 6.1	Students make a plan to find how long it would take to tap on a desk a million times.	What units are important to consider when making this plan? How large a sample will you need in order to get useful information to solve the problem? **What information do you need to use in order to solve your problem? What measurement conversions will you need to make?** Why is it important to make a plan before you begin solving a problem?

Grade 6 Unit 1
Collection, Display, and Interpretation of Data

Activity	*Everyday Mathematics* Goal for Mathematical Practice	Opportunity	Guiding Questions
Lesson 1♦1: Introduction to the *Student Reference Book*			
Using the *Student Reference Book* (*Teacher's Lesson Guide,* pages 19 and 20)	**GMP 5.2** Use mathematical tools correctly and efficiently. *See also:* **GMP 3.2, GMP 6.3**	Students use the *Student Reference Book* to help them solve problems and answer questions about various mathematical topics.	How did you find helpful information in the *Student Reference Book* for each question? What can you do to make sure you understand the information in the *Student Reference Book*? How can the *Student Reference Book* help you in class or at home? **How can tools help you to be a better mathematician?**
Lesson 1♦2: Line Plots			
Matching Line Plots with Statements and Landmarks (*Teacher's Lesson Guide,* pages 24 and 25)	**GMP 2.2** Explain the meanings of the numbers, words, pictures, symbols, gestures, tables, graphs, and concrete objects you and others use. *See also:* **GMP 1.6, GMP 2.1, GMP 3.1, GMP 4.1, GMP 6.1**	Students match "mystery" line plots that display data about themselves with statements about the data.	What information in the line plots did you use to match the line plots with the statements? What conclusions can you draw about our class from the information in the line plots? Using the data displayed in the line plots, what predictions can you make about another sixth-grade class? **Why is it important to understand what graphs mean?**
Lesson 1♦3: Stem-and-Leaf Plots			
Math Message Follow-Up (*Teacher's Lesson Guide,* pages 28 and 29)	**GMP 1.6** Connect mathematical ideas and representations to one another. *See also:* **GMP 4.1, GMP 6.1**	Students compare and contrast four different representations of the same test-score data.	How does each representation show information about the test scores? How are the representations alike? How are they different? What are the advantages and disadvantages of each representation of the test-score data? **Why is it important to be able to represent data in more than one way?**

Grade 6 Unit 1
Collection, Display, and Interpretation of Data (cont.)

Activity	*Everyday Mathematics* Goal for Mathematical Practice	Opportunity	Guiding Questions
Lesson 1♦4: Median and Mean			
Comparing the Median and Mean of a Data Set (*Teacher's Lesson Guide*, page 35)	**GMP 4.1** Apply mathematical ideas to real-world situations. See also: **GMP 2.2, GMP 6.1**	Students consider the use of median and mean as they apply to test scores.	What information does the mean give you about the test scores? What information does the median give you about the test scores? What are other real-world situations in which knowing the median and mean of a data set is useful? **How can mathematics help you make decisions in the real world?**
Lesson 1♦5: Playing *Landmark Shark*			
Playing *Landmark Shark* (*Teacher's Lesson Guide*, page 40)	**GMP 3.1** Explain both what to do and why it works. See also: **GMP 3.2, GMP 6.1**	After playing several rounds of the game *Landmark Shark*, students discuss their strategies for getting high scores.	Explain your strategies for getting high scores while playing *Landmark Shark*. Did you ever exchange cards? Why or why not? Do you think *Landmark Shark* is more a game of luck or a game of strategy? Why? **Why is it important to be able to explain a game strategy?**
Lesson 1♦5a: Box Plots			
Constructing Box Plots: Hair Length by Gender (*Teacher's Lesson Guide*, page 42D)	**GMP 2.1** Represent problems and situations mathematically with numbers, words, pictures, symbols, gestures, tables, graphs, and concrete objects. See also: **GMP 1.6, GMP 2.2, GMP 4.1, GMP 6.1**	Students draw two box plots that show the distribution of the hair lengths of boys and girls in the class.	How do the box plots represent the hair-length data you collected? Use the information shown by the box plots to describe the distribution of the girls' hair lengths and the boys' hair lengths. What other ways could you organize this data to help you analyze it? **Why is it important to organize data?**

Grade 6 Unit 1
Collection, Display, and Interpretation of Data (cont.)

Activity	*Everyday Mathematics* Goal for Mathematical Practice	Opportunity	Guiding Questions
Lesson 1◆6: Broken-Line Graphs			
Drawing and Interpreting a Broken-Line Graph (*Teacher's Lesson Guide*, pages 44 and 45)	**GMP 2.2** Explain the meanings of the numbers, words, pictures, symbols, gestures, tables, graphs, and concrete objects you and others use. **See also:** GMP 1.6, GMP 2.1, GMP 4.1	Students construct a broken-line graph and discuss the information displayed in the graph.	What features of the line graph help you understand the information it represents? What does the shape of the graph tell you about precipitation in Omaha throughout the year? What predictions about precipitation in Omaha can you make using the graph? **How is it helpful to see data displayed in a graph?**
Lesson 1◆7: Bar Graphs			
Math Message Follow-up (*Teacher's Lesson Guide*, pages 49 and 50)	**GMP 2.1** Represent problems and situations mathematically with numbers, words, pictures, symbols, gestures, tables, graphs, and concrete objects. **See also:** GMP 2.2, GMP 4.1, GMP 6.1	Students construct a bar graph using information about students' scores on a math quiz.	How did you decide what labels to give the horizontal and vertical axes? Why is it important to title your graph and label the axes? **How does organizing information in a graph help you understand the information better?**
Lesson 1◆8: Step Graphs			
Playing *Name That Number* (*Teacher's Lesson Guide*, page 57)	**GMP 6.3** Be accurate when you count, measure, and calculate. **See also:** GMP 1.3, GMP 6.1	Students write number sentences using order of operations as they play the game *Name That Number*.	How did you make sure your solutions were correct while you played? What did you do to check your partner's solutions? What did you need to remember about the order of operations while you were writing your solutions? **What are good strategies for making sure your answers are correct?**

Grade 6 Unit 1
Collection, Display, and Interpretation of Data (cont.)

Activity	*Everyday Mathematics* Goal for Mathematical Practice	Opportunity	Guiding Questions
Lesson 1◆9: The Percent Circle and Circle Graphs			
Practicing with the Percent Circle (*Teacher's Lesson Guide,* pages 62 and 63)	**GMP 5.2** Use mathematical tools correctly and efficiently. *See also:* **GMP 2.2, GMP 4.1, GMP 5.3**	Students answer questions about information in circle graphs using a Percent Circle.	How did the Percent Circle help you answer the questions about the circle graphs? Did you have any difficulties using the Percent Circle? If so, what did you do about them? What do you need to remember to use the Percent Circle correctly? **Why is it important to use mathematical tools correctly?**
Lesson 1◆10: Using a Graph to Investigate Perimeter and Area			
Finding the Largest Area for a Given Perimeter (*Teacher's Lesson Guide,* pages 67 and 68)	**GMP 4.2** Use mathematical models such as graphs, drawings, tables, symbols, numbers, and diagrams to solve problems. *See also:* **GMP 2.1, GMP 4.1, GMP 8.1**	Students use a graph to solve a problem involving the relationship between the area and perimeter of rectangles.	How does the graph show the relationship between the area and length of the rectangles? How can you use the graph to find the largest area enclosed by 22 feet of fence? What is another problem that could be solved using this graph? **How do graphs and other mathematical models help you solve problems?**
Lesson 1◆11: Persuasive Data and Graphs			
Analyzing a Persuasive Pictograph (*Teacher's Lesson Guide,* pages 73 and 74)	**GMP 3.2** Work to make sense of others' mathematical thinking. *See also:* **GMP 2.2, GMP 4.1, GMP 6.1**	Students analyze a misleading pictograph to find errors in the representation.	What message is the author of this graph trying to send? What questions would you ask the author about the graph? How could you represent the message of the graph more accurately? **Why is it important to be able to make sense of others' mathematical thinking?**

Grade 6 Unit 1
Collection, Display, and Interpretation of Data (cont.)

Activity	Everyday Mathematics Goal for Mathematical Practice	Opportunity	Guiding Questions
Lesson 1♦12: Samples and Surveys			
Math Message Follow-Up (*Teacher's Lesson Guide,* pages 77 and 78)	**6.1** Communicate your mathematical thinking clearly and precisely. *See also:* **GMP 3.1, GMP 3.2, GMP 6.1**	Students explain how to find the percentage of the cookie weight that is chocolate chips.	How did you and your partner explain your thinking so that you understood each other? Show or tell how you would solve the problem so that someone else could use your method. What numbers, units, symbols, labels, or drawings can you use to make sure your thinking is clear to others? **Why is it important to explain your thinking clearly as well as to give a correct answer?**

Grade 6 Unit 2
Operations with Whole Numbers and Decimals

Activity	*Everyday Mathematics* Goal for Mathematical Practice	Opportunity	Guiding Questions
Lesson 2•1: Reading and Writing Large Numbers			
Interpreting Expanded Notation for Large Numbers (*Teacher's Lesson Guide*, page 104)	**GMP 1.6** Connect mathematical ideas and representations to one another. *See also:* **GMP 2.2, GMP 5.2, GMP 7.1**	Students use a place-value chart to convert numbers from standard notation to expanded notation.	How can you use extended multiplication facts to write large numbers in expanded notation? What does expanded notation tell you about a number that standard notation does not? Compare standard notation and expanded notation. How are they alike and different? **How are different representations of the same number useful?**
Lesson 2•2: Reading and Writing Small Numbers			
Math Message Follow-Up (*Teacher's Lesson Guide*, pages 109 and 110)	**GMP 7.1** Find, extend, analyze, and create patterns. *See also:* **GMP 1.6, GMP 2.2, GMP 6.1**	Students identify and describe patterns in a place-value chart.	What patterns do you see in the place-value chart? Look at the place-value chart on journal page 45. How do the patterns on these two charts compare? Why do you think that is? **How can just ten digits form numbers that are greater than one and numbers that are less than one?**
Lesson 2•3: Addition and Subtraction of Decimals			
Drawing and Interpreting Histograms (*Teacher's Lesson Guide*, page 116)	**GMP 4.1** Apply mathematical ideas to real-world situations. *See also:* **GMP 2.1, GMP 2.2, GMP 4.2**	Students compare two histograms that display the same data with different intervals on the horizontal axes.	Which histogram gives more information about the students' reading habits? Why do you think so? What questions about the reading habits of the class can you answer using these data? Is there another way to display the data that would give more information about the students' reading habits? **How is it helpful to use displays like histograms to represent real-world information?**

Grade 6 Unit 2
Operations with Whole Numbers and Decimals (cont.)

Activity	*Everyday Mathematics* Goal for Mathematical Practice	Opportunity	Guiding Questions
Lesson 2◆4: Multiplying by Powers of 10			
Math Message Follow-Up (*Teacher's Lesson Guide,* pages 119 and 120)	**GMP 8.1** Use patterns and structures to create and explain rules and shortcuts. See also: GMP 1.6, GMP 5.3, GMP 6.1, GMP 7.1	Students find patterns in a table that shows powers of 10 represented in exponential notation, standard notation, and repeated multiplication. They use the patterns to develop shortcuts for multiplying by powers of 10.	Describe the patterns you see in the rows and the columns of the table. How are these patterns connected to place value? Use the patterns you noticed to explain the strategies you found for multiplying by powers of 10. **How does studying patterns lead to mathematical shortcuts?**
Lesson 2◆5: Multiplication of Decimals: Part 1			
Estimating Products of Decimals (*Teacher's Lesson Guide,* page 125)	**GMP 3.1** Explain both what to do and why it works. See also: GMP 3.2, GMP 5.3, GMP 6.1	Students estimate products of decimals and explain their strategies.	Explain your estimation strategy so someone else could use it. Why does your estimation strategy work? What words, numbers, tools, or drawings can you use to make your explanation clear? **Why is it important to explain why your strategies work?**
Lesson 2◆6: Multiplication of Decimals: Part 2			
Math Message Follow-Up (*Teacher's Lesson Guide,* pages 130 and 131)	**GMP 6.3** Be accurate when you count, measure, and calculate. See also: GMP 3.2, GMP 5.3, GMP 6.1	Students use lattice multiplication to multiply decimals.	What do you need to remember to be sure you are accurate when you use lattice multiplication with decimals? How confident are you that your answer is correct? Explain why. **What can you do before, during, and after you multiply to be sure your answer is correct?**

Grade 6 Unit 2
Operations with Whole Numbers and Decimals (cont.)

Activity	*Everyday Mathematics* Goal for Mathematical Practice	Opportunity	Guiding Questions
Lesson 2◆7: Division of Whole Numbers			
Using the Partial-Quotients Division Algorithm (*Teacher's Lesson Guide,* page 139)	**GMP 6.3** Be accurate when you count, measure, and calculate. See also: **GMP 5.3**	Students solve division problems using the partial-quotients algorithm.	Explain your strategies for finding partial quotients. Can you reduce the number of steps to solve a division problem with partial quotients and still be accurate? How? How confident are you that your answer is correct? Explain why. **What can you do before, during, and after you divide to be sure your answer is correct?**
Lesson 2◆8: Division of Decimals			
Interpreting Remainders (*Teacher's Lesson Guide,* page 145)	**GMP 1.5** Check whether your solution makes sense. See also: **GMP 4.1, GMP 6.2, GMP 6.3**	Students solve division number stories and interpret the remainders based on the context of the number story.	How did you decide what to do with the remainders as you solved the problems? **Why is it important to look back at the original problem to be sure your solution makes sense?** How does checking whether your answer makes sense help you solve problems?
Lesson 2◆9: Scientific Notation for Large and Small Numbers			
Translating between Scientific and Standard Notations (*Teacher's Lesson Guide,* page 149)	**GMP 1.6** Connect mathematical ideas and representations to one another. See also: **GMP 8.2**	Students translate large and small numbers between scientific and standard notations.	How are standard notation and scientific notation alike and different? What are the advantages and disadvantages of each notation? When might it be better to use standard notation? When might it be better to use scientific notation? **Why is it important to be able to represent the same number in more than one way?**

Grade 6 Unit 2
Operations with Whole Numbers and Decimals (cont.)

Activity	*Everyday Mathematics* Goal for Mathematical Practice	Opportunity	Guiding Questions
Lesson 2◆10: Exponential Notation and the Power Key on a Calculator			
Working with Exponents on a Calculator (*Teacher's Lesson Guide*, page 154)	**GMP 5.2** Use mathematical tools correctly and efficiently. See also: **GMP 1.6, GMP 7.1**	Students use calculators to convert numbers written in exponential notation to standard notation.	What do you know about exponents that helps you convert numbers from exponential to standard notation with a calculator? Describe two ways to use a calculator to convert numbers from exponential to standard notation. Which way is more efficient? **When is it a good idea to use tools like a calculator?**
Lesson 2◆11: Scientific Notation on a Calculator			
Using Scientific Notation on a Calculator (*Teacher's Lesson Guide*, page 159)	**GMP 3.2** Work to make sense of others' mathematical thinking. See also: **GMP 5.2**	Students follow examples in the *Student Reference Book* to learn how to work with numbers represented in scientific notation on a calculator.	Which example was the easiest to follow and understand? Why? Which example was the hardest to follow and understand? Why? What can you do to be sure you can follow the steps the author shows in the examples in the *Student Reference Book*? **How does making sense of someone else's mathematical thinking help your mathematical thinking?**